THE GIFT OF MUSIC—FOR CHILDREN AND THOSE WHO LOVE THEM

The Mozart Effect® for Children

is an essential guide that helps parents and teachers unlock a child's intelligence and creative potential and shows how music, sound, and tone

- Influence babies whose mothers had prenatal exposure to music, resulting in better visual tracking, eye-hand coordination, and other positive behaviors.

- Strengthen the child's body, even prior to birth. In fact, studies show that the unborn baby "feels" sound—including and especially the sound of your voice—for months before she ever hears it.

- Stimulate a child's brain while nourishing the spirit and encouraging mental awareness and creative discovery.

- Improve the child's language, movement, and emotional skills and help him reach peak emotional and intellectual potential.

- Help regulate both the child's physical activity and the give-and-take of social interaction.

- Bolster a child's sense of personal identity and awareness of environment.

- Encourage articulate speech.

- Help develop the child's creativity, allowing her to discover that creativity enters into every branch of life: relationships, conversation, and academic work.

"Love, love, love, that is the soul of genius."

—Wolfgang Amadeus Mozart

David Mays

About the Author

DON CAMPBELL has spent more than three decades exploring and writing about the benefits of music for lifelong learning, and is a world-renowned authority on music and its role in education and health. A classical musician, educator, writer, and teacher, he works with symphony orchestras, schools, music educators, and health professionals worldwide. He has helped children in thirty countries improve their abilities to learn, create, and experience the joy of life through music. He serves on the boards of the Boulder Philharmonic and the American Music Research Center. His books have been translated into a dozen languages. Mr. Campbell lives in Boulder, Colorado.

BOOKS AND TAPES BY DON CAMPBELL

Books

Music and Miracles
100 Ways to Improve Teaching Using Your Voice & Music
Music: Physician for Times to Come
Rhythms of Learning (with Chris Brewer)
The Roar of Silence
Master Teacher: Nadia Boulanger
Introduction to the Musical Brain
The Mozart Effect®

Spoken Tapes

The Power of Music (five-tape set)
Mozart as Healer
Healing Yourself with Your Own Voice
Healing with Great Music
Healing with Tone and Chant
Sound Pathways: Accelerated Learning and Music

Music Tapes and Compact Discs

Essence

Music for the Mozart Effect®
Vol. I—Strengthen the Mind: Music for Intelligence and Learning
Vol. II—Heal the Body: Music for Rest and Relaxation
Vol. III—Unlock the Creative Spirit: Music for Creativity and Imagination
Vol. IV—Music for Stress Reduction
Vol. V—Music for Study

The Mozart Effect®—Music for Children
Vol. I—Tune Up Your Mind
Vol. II—Relax, Daydream, and Draw
Vol. III—Mozart in Motion
Vol. IV—Mozart to Go!

The Mozart Effect®—Music for Babies
Vol. I—From Playtime to Sleepytime
Vol. II—Music for Newborns
Vol. III—Nighty-Night Mozart

The Mozart Effect®—Music for Moms and Moms-to-Be

Audio Books

The Roar of Silence
The Mozart Effect®

The Mozart Effect® for Children

Awakening Your Child's Mind, Health, and Creativity with Music

Don Campbell

Quill
An Imprint of HarperCollins*Publishers*

To

the students of St. Mary's International School

in Tokyo, Japan,

the children at Grace Children's Hospital

in Port-au-Prince, Haiti,

and

the teachers at the Guggenheim School

in Chicago, Illinois

Permission to reprint the excerpt from Judith Morley's *Miss Laughinghouse and the Reluctant Mystic* granted by Black Thistle Press, copyright © 1995.

The Mozart Effect® is the registered trademark of Don G. Campbell, Inc., and is used with permission throughout this book.

A hardcover edition of this book was published in 2000 by William Morrow, an imprint of HarperCollins Publishers.

HarperCollins books may be purchased for educational, business, or sales promotional use. For information please write: Special Markets Department, HarperCollins Publishers Inc., 10 East 53rd Street, New York, NY 10022.

First Quill edition published 2002.

Designed by Nancy Singer Olaguera

Library of Congress Cataloging-in-Publication Data is available.

ISBN 0-380-80744-0

02 03 04 05 06 RRD 10 9 8 7 6 5 4 3 2 1

CONTENTS

ACKNOWLEDGMENTS

I am especially grateful to Michelle Henderson, president of The Children's Group, for her determination and business acumen in marketing the children's Mozart series, which helped pave the way for this book.

A huge thanks to Sherill Tippins for her ability to translate my years of research and thoughts into a fresh form, and for her valuable input regarding early childhood education.

Deep appreciation goes to my supportive and innovative agent, Eileen Cope, who has been with me from the very beginning. I would also like to thank Barbara Lowenstein, as well as Tom Mone and Norman Kurz.

Special thanks to my manager, Bill Horwedel, who has made an enormous impact and contribution to my books and CDs and who has brilliantly coordinated my career. Thanks also to his assistant, Noelle Cangialossi.

I am very grateful to my first editor, Rachel Klayman, for her editorial genius; my new editors, AnnMcKay Thoroman and Jennifer Brehl, for jumping in with tremendous enthusiasm; and the rest of the staff at Morrow/Avon, particularly Lou Aronica, Michael Greenstein, Lisa Queen, and Brett Witter.

Thanks also to:

Kellie Masterson, Dee Coulter, and Carla Hannaford, who assisted in the accuracy of the research and details.

Sebastien, Elizabeth, Victoria, and Ann, whom I have known since before their births, and who allowed me to witness their growth in brain, body, and heart.

Charles Fowler for his visionary inspiration and advice. Dr. Alfred Tomatis, Paul Madaule, Billie Thompson, Ph.D., Judith Belk, Ph.D., Pat Cook, Ph.D., and Pierre Sollier for their innovative and visionary research in sound. Norm Goldberg, Marcia Goldberg, Douglas Jones, and Naomi Tobias for their constant support.

Brad Butler, Sue Pfeifer, Bea Romer, and the visionary staff of Bright Beginnings, who help newborns into harmony throughout Colorado.

The dozens of associates who have shared magical stories about music and children: Eileen Auxier, Joanne Bath, Barbara Boulton, Alice Cash, Carla Crutsinger, Debbie Fier, Patricia Smith, Richard Grossheider, Dorothy Jones, Linda Kittchner, Deborah Klenke, Carol Kranowitz, Marilyn Larson, Heather Lloyd, John Ortiz, Nick Page, Janice Reade, Nyansa Susan Tobin, Vicki Vorreiter, Aubrey Carton, Sara Cakebread.

Kenneth K. Guilmartin, Liz Gilpatrick, Judith Cole, Jane Smolens, Kathy Kucsan, Diane Howard, and the teachers at Douglass Elementary School in Boulder. Karen Romero of the Boulder Academy, Chris Brewer, Martha Hallquist, Grace Nash, Lorna Lutz Heyge, and the many researchers and teachers who have devoted their lives to music in early childhood.

And special thanks to the students and parents of Bohemia Elementary School in Cottage Grove, Oregon, who are providing new models for keeping music alive in their school.

FOREWORD

The Mozart Effect for Children is a major contribution to a quiet revolution that is under way today, one that could change human history as decisively as printing, electronics, or quantum physics. And for the better. This gentle revolution is arising from research about the formation and development of a child's brain from conception. No generation before us has had access to such a wealth of knowledge about infant-child development as we have today, nor as clear directives for how to apply the information to ourselves and our children. The difference such an application can make in a human life is astonishing, indeed revolutionary.

Thanks to the devotion and skill of scholars such as Don Campbell, the future of humankind has never looked so bright. In the pages that follow, Campbell offers us a simple yet comprehensive summary of child development, centering on his broad grasp of the brain research taking place in our time. Equally he offers a novel, creative analysis of how we can apply this research in our own lives, as well as those of our children. I know of no other work that has managed in one short volume to make so complex a subject so clear, engrossing, delightful, and easy to read, as well as giving such clear directives for how to apply and make immediate use of what is known. Campbell's is a rare gift, and this is a rare work.

I learned so much from this book, in fact, and on a subject I thought I was fairly proficient in, that I feel genuine gratitude for it. I had a similar experience when, having written three books centering on the intelligence of the heart, I discovered the Institute of Heartmath and the new medical field of neurocardiology. Neurocardiology translates, roughly, as "the brain in the heart" and centers on the discovery that the heart is a major neural center within us and the source of the actual "governing intelligence" in our life. Research has lifted the issue of the heart out of its ancient category of posy and sentiment into an astonishing biological factor critical to human intelligence. In the same way, Don Campbell, through his many decades of research into music and child development, has lifted music out of its category of ornament, frill, or entertainment into its rightful place as one of the major innate intelligences. Genetically encoded into our species and vital to our well-being, music is, indeed, a critical foundation for most of our higher intelligences, as Campbell amply shows.

Howard Gardner listed music as one of the seven basic intelligences built into our genetic system. Even earlier, following Maria Montessori, Jean Piaget recognized music as an innate intelligence ready to unfold between the third and fourth year of life. Earlier still, through his intuitive grasp of human development in general, Rudolf Steiner made music a cornerstone of his remarkable Waldorf education. Steiner recognized music to be foundational to intellect, creativity, mathematical capacity, and spiritual development, as well as perhaps the greatest of art forms in its own right.

A lifelong music lover, I recall reading with pleasure, back in the late 1940s, that the atomic scientists at Oak Ridge, Tennessee, gathered at spare moments in the evening to play string quartets. I was intrigued that when Nobel laureate David Hubel, a neuroscientist, was asked whether he had any interests other than his specialty—research into the visual brain—he replied, "Actually I seem to have spent an inordinate amount of my life at the piano." And, of course, Einstein's long love affair with the violin is well known.

Don Campbell throws light on the role of sound itself in body-brain process, calling on his long association with the French physician Alfred Tomatis, the world's foremost authority on the role sound plays in consciousness and perception. Campbell has, as well, a remarkable understanding and knowledge of infant-child development, having worked in depth with children at all levels. His view of child development as it relates to music presents an aspect of intellectual growth not approached at such depth by anyone before this time (of which I am aware). And surely, from Campbell's extraordinary breadth of data, the reader will see why music emerges as a major factor in the growth of intelligence in general.

Interestingly, ancient Hindu scholars claimed the universe arose first as sound; sound gave rise to light, and light to matter. As with the

Bible, eastern sages stated that in the beginning was the Word, the sound that was with, and/or in, God. We know that in the development of the uterine brain-mind, the auditory system forms very early. And language learning, in its primary sensory-motor form, begins around the fifth uterine month, when the infant begins making muscular responses to the phonemes (or "phonics," if you like) of the mother's language. And this bodily response to sound and word precedes and is foundational to development of the visual word. Vision must await birth for its unfolding, of course, and on the opening of a "visual world" the infant-child builds those "structures of knowledge" concerning the physical world experienced through the early years. So the ancients were correct in their sequencing of sound, word, light, and matter as the way our human world unfolds for us.

Physicist David Bohm, protégé of Einstein, spoke of matter as frozen light and music as "pure implicate order," the implicate order being the subtle quantum energy from which, according to Bohm, all experience springs. Philosopher Susanne Langer proposed that language arose from singing, and that singing arose from spontaneous expressions of ecstasy or joy. And Plato said that if he could choose the music young people listened to and performed, he could determine the society they would bring about.

I noted with admiration (and some mild envy) the efficient threefold "teaching procedure" Campbell uses here, and how carefully he has worked to make our journey through this story of development a delightful esthetic experience as well as an adventure of learning. At the same time he gives us rich, pertinent examples that illustrate and make clear the developmental issues at stake. Third, he offers orderly ways we can apply each level of learning he covers, in concrete and eminently practical steps any of us can follow, to make real in our child's life, as well as our own, the vast richness of spirit offered here.

Campbell is well aware of the severe difficulties children face today—the mounting abandonment in infancy, the lack of nurturance, the inadequacies of education. But rather than dwelling on these negatives, he keeps our energy and attention focused on doing what is constructive and beneficial. In a splendid example of what I would call the "model imperative," Campbell offers biographical glimpses into the nature of music's great genius, Mozart. This genius, Campbell makes clear, was brought out and nurtured to its full maturation by parents who gave the necessary nurturing, stimulus, encouragement, and support for that intelligence to flower. This book as a whole seems designed, in fact, to help us bring out the "Mozart" in each of our children, in whatever medium they might be predisposed to express their genius.

I relish for the reader the delight and enlightenment offered in the following, a marvelous course in child and human development from a perspective so new, so fresh, that you can't help but be enriched on many levels. Were all that not enough, this book is an introductory course to music listening and appreciation, a young *or* old person's guide to the greatest of great classics. So whether you are a parent with an infant, middle child, or adolescent; childless, or intend to have a child someday; or, perhaps, are a child at heart (or have a perpetual child as a spouse)—whatever your situation, the following is a rich and rewarding experience.

—Joseph Chilton Pearce, author of
Magical Child and *Evolution's End: Claiming the Potential of Our Intelligence*

PRELUDE

How Magic Is Thy Sound

Neither a lofty degree of intelligence nor imagination
nor both go to the making of genius.
Love, love, love, that is the soul of genius.
—WOLFGANG AMADEUS MOZART

My musical odyssey started before I was born, when my father's joyful harmonica, guitar, and piano playing first reached my developing ears. It gathered momentum in my early childhood as I listened, entranced, to my 78 rpm of Daffy Duck singing silly lyrics to Liszt's *Hungarian Rhapsody*. I can still remember the movements of my three-year-old body as I gestured to "I'm a Little Teapot" in Sunday school class. I can still feel the hard gym floor where I rested after lunch while Sister Mary played "The Laughing Phonograph." I can smell the furniture polish on the dining table in our house as I sang along with a recording of "Dites Moi" from *South Pacific*, finding meaning in words I didn't understand.

Everything was alive with a rhythm I knew nothing about yet understood intimately and thoroughly. I remember singing, singing, singing to doodlebugs, trees, and to my imaginary friend, Hector Hamhock. The sounds of saws and hammers in my father's workshop thrilled with their sharp, bright tones. Music was the pleasure, the power that ruled my days. Dad walked and whistled, Mom's food sizzled on the stove, my grandmother Mimi's rocker creaked, and Perfume, the cat, purred under my touch. It was all music.

My childhood love affair with sound led to a long and fulfilling career in music: composing and performing professionally, educating children through rhythm and tone, and investigating the many ways in which music shapes and stimulates the mind, body, and spirit. Studying as a boy with Nadia Boulanger on the Fontainebleau Conservatory of Music near Paris; teaching music and philosophy at St. Mary's International School in Tokyo; observing musical rite-of-passage rituals in Bali; and singing to and helping young tuberculosis patients communicate in Haiti, I was able to compare the music of different cultures and its effects on people everywhere. While taking part in the Guggenheim Education Project in Chicago's inner city, founding and operating the Institute for Music, Health, and Education in Boulder, Colorado, and consulting for numerous educational organizations and schools, I had the opportunity to study all the ways in which rhythm, tone, and other aspects of music and sound can bolster creativity, stimulate the mind, heal the body, vanquish stress, and strengthen families.

In the years that followed, I wrote a number of books on music and education, the brain, and the spirit, and began leading workshops around the world to spread the word about the power of sound. My fascination grew intensely personal in 1994, when I was diagnosed with a large blood clot just beneath the right hemisphere of my brain. If I truly believed that music could physically affect the mind and body, now was certainly the time to put my belief to the test. Over the months that followed, I learned to use sound and mental imagining to slow down my pulse, breathing, and metabolism, to release energy to flow through my mind and body, and thus help ease my psychological concerns and correct my physiological state. The disappearance of the blood clot increased my great respect for the power of sound and inspired me to write *The Mozart Effect*.

Certainly I was not the first person intrigued by the mysterious effects of music—or Mozart. I first encountered a reference to Mozart's extramusical potential as a teenager, watching Alfred Hitchcock's 1957 movie, *Vertigo*. In the film, Midge Wood (Barbara Bel Geddes) remarks to John "Scottie" Ferguson (James Stewart), who lies depressed in his hospital bed, "I had a long talk with a lady in musical therapy. She says Mozart's the boy for you . . . the one that sweeps the cobwebs away." Interestingly, *Vertigo* was inspired by a French novel, and it was in France that the greatest crusader in the area of sound's effect on brain and body development—Dr. Alfred Tomatis—began using Mozart's music in his research in the 1960s and 1970s. Tomatis was able to prove that listening to certain filtered sounds—specifically the sounds of Mozart's music or the sound of the mother's voice—actually affects the brain in a way that improves listening and speech skills, emotional health, and mental alertness. His highly successful work with children suffering from attention deficit disorder and other learning problems proved particularly exciting. News of his accomplishments joined a rising tide of scientific reports verifying an array of demonstrable, measurable, positive effects of music.

During the 1980s and 1990s, scientific journals around the world began publishing studies proving that music literally alters the structure of the developing brain of the fetus; that infants recognize and prefer music first heard in their mothers' wombs; that IQ scores increase among young children who receive regular music instruction; that a single half hour of music therapy improves children's immune function; and that music relieves stress, encourages social interaction, stimulates language development, and improves motor skills among young children. Finally, educators, musicians, and scientists began to ask the right questions: How does music affect complex systems in the brain and body? How can music increase memory, decrease stress, and enhance physical performance? Why does music create different responses, depending on whether the listener is young or old, sits slumped or straight, listens in the morning or at night?

During the past decade, I have worked to spread the word about music's power to such varied groups as symphony orchestras, the La Leche League, the Family Therapy Network, and many others. Wherever I have spoken, I have been approached by those who had experienced their own transformations through music, or who were intrigued by the positive effects music seemed to have on their children. "Is it true that playing the piano makes kids smarter?" call-in radio listeners would ask. "Can music improve my son's sports ability?" "My daughter is so shy. Can music help her make friends?" "Can music help my baby learn to walk sooner?" "Can it make the birthing process easier?" "Can it help my son learn his multiplication tables?" "Is it true that music actually changes the structure of the brain?"

The Mozart Effect for Children is designed to meet this demand for more information about the valuable extramusical properties of music and sound—specifically about the many ways in which rhythm and tone can enhance your child's development from before birth into middle childhood and beyond. In this book, we will trace the growth of the child as she develops a brain and sensory organs before birth, is born into a bewildering but utterly captivating new world, and gradually learns to organize, manipulate, and master all she sees. We will learn how the familiar tones of her mother's voice and the comforting rhythm of the maternal heartbeat can soothe her before, during, and after the birth process; how music's varying pitch can awaken her ears and stimulate language development; and how rhythmic games and nursery chants can teach her body to move with grace and ease.

As the child grows, her social and academic skills can be enhanced by her relationship with music. Music can mirror her half-understood emotions and help her learn to express what she feels. Making music with others can strengthen her bonds with her family and community and connect her to her cultural heritage.

Throughout this book, I will provide specific exercises to help your child reap the benefits music offers. I will introduce you to some of the hundreds of thinkers and educators who are adding daily to our store of knowledge in this dynamic new field. (For those parents and educators who wish to learn more, additional notes are included at the end of the book.) I will offer examples of the many ways music has improved the lives of children with learning disabilities, emotional disorders, and physical challenges. I will explore ways in which music could improve our elementary education system. In short, I will show you how simply music—that miraculous force that is available to all of us free of charge—can help your tiny infant grow up to become a confident, thoughtful, well-balanced child.

Can music make your child more intelligent? Certainly it can increase the number of neuronal connections in her brain, thereby stimulating her verbal skills. It can teach her good study habits, aid in her efforts to read and to comprehend mathematical concepts, and help her memorize facts with ease. But intelligence is not measured only by our ability to read, write, memorize, and work with numbers. Our success in working in community, in remembering visually and aurally, in moving, creating, and interacting with grace and sensitivity, in expressing emotion and relieving stress, and in listening to and trusting our own "inner voice" are equally important—and all are enhanced by listening to and making music. True, many influences contribute to the molding of a life, and music is only one of them. But unlike our genetic inheritance, which is fixed, our musical inheritance is expandable. We can turn up the volume and make it as positive a force as we wish.

In short, my intention in this book is not to create superprodigies by showing you how to stuff your child's head full of information using musical techniques. My goal is simply to give as many children as possible the incomparable gift of music—and, in doing so, to help them reach their emotional, intellectual, and spiritual potential. Of course, I would also like to see growing children appreciate and create great music purely for the sake of enjoyment and aesthetic pleasure. Parents and children should not feel that music must always be useful; often it is simply beautiful, and that is more than enough. However, my mission—and the message of this book—goes beyond enjoying a beautiful song or creating a proficient musician, valuable as I believe those goals to be. It is about the extramusical benefits of sound. It is about transforming and enriching children's lives, and ours at the same time.

In the end, music's greatest power may lie in its ability to capture the joy you feel every day with your little one, and to contain both of you in its embrace. Researchers will continue to probe music's mysteries for centuries to come, but they will never be able to isolate and quantify the magical effects of a spontaneous song or dance shared by parent and child. These enchanting moments, amplified by love, are, as Mozart's music reminds us, the truest and most important song of life.

CHAPTER 1

TWINKLE TWINKLE, LITTLE NEURON

Music and Your Child's Brain

Ah! Vous dirai-je, Maman,	Ah! Let me tell you, Mother,
Ce qui cause mon tourment?	What's the cause of my torment?
Papa veut que je raisonne,	Papa wants me to reason
Comme une grande personne;	Like a grown-up.
Moi, je dis que les bonbons	Me, I say that candy has
Valent mieux que la raison.	Greater value than reason.

—EIGHTEENTH-CENTURY FRENCH FOLK SONG

Long before the lyrics to "Twinkle Twinkle, Little Star" were written, children across France sang the words you see above to the same tune. Seventeen-year-old Wolfgang Amadeus Mozart must also have been familiar with the song, since he used its melody as a starting point for his playful, ever expanding *Variations on Ah! Vous dirai-je, Maman* (K. 265). Might the brilliant teenager have chosen this melody to tease his notoriously stern, ambitious father, Leopold, for his taskmaster approach toward raising a son? Given Wolfgang's love of jokes and clever wordplay, it certainly seems likely.

More important, though, Mozart's *Variations*, now practiced and memorized by intermediate music students around the world, perfectly evoke the way we humans best think and grow creatively. After all, as Mozart might tell us if he were alive today, pleasing, organized melodies such as this one do have great value, particularly for children. Music speaks in a language that children instinctively understand. It draws children (as well as adults) into its orbit, inviting them to match its pitches, incorporate its lyrics, move to its beat, and explore its emotional and harmonic dimensions in all their beauty and depth. Meanwhile, its physical vibrations, organized patterns, engaging rhythms, and subtle variations interact with the mind and body in manifold ways, naturally altering the brain in a manner that one-dimensioned rote learning cannot. Children are happy when they are bouncing, dancing, clapping, and singing with someone they trust and love. Even as music delights and entertains them, it helps mold their mental, emotional, social, and physical development—and gives them the enthusiasm and the skills they need to begin to teach themselves.

In recent decades, an enormous amount of research has been conducted on the specific ways in which sound, rhythm, and music can improve our lives. The results of the research using Mozart's music have been especially stunning and have given rise to the term the Mozart Effect. I use the phrase to encompass such phenomena as the ability of Mozart's music to temporarily heighten spatial awareness and intelligence; its power to improve listeners' concentration and speech abilities; its tendency to advance the jump in reading and language skills among children who receive regular music instruction; and the startling increase in SAT scores among students who sing or play an instrument. But the Mozart Effect refers to more than just raising children's test scores. By learning to recognize and consciously implement the Mozart Effect in your child's life, you can:

- Begin to communicate and connect with him even before he is born.
- Stimulate brain growth in the womb and throughout early childhood.
- Positively affect his emotional perceptions and attitudes from prebirth onward.
- Provide patterns of sound on which he can build his understanding of the physical world.

- Reduce his level of emotional stress or physical pain, even in infancy.
- Enhance his motor development, including the grace and ease with which he learns to crawl, walk, skip, and run.
- Improve his language ability, including vocabulary, expressiveness, and ease of communication.
- Introduce him to a wider world of emotional expression, creativity, and aesthetic beauty.
- Enhance his social abilities.
- Improve his reading, writing, mathematical, and other academic skills, as well as his ability to remember and to memorize.
- Introduce him to the joys of community.
- Help him create a strong sense of his own identity.

It is amazing to think that music and rhythmic verbal sounds, which have been available to us throughout our lives, can have such a powerful effect on the mind and body. Yet the evidence is indisputable. There's far more to good music than meets the ear. Wisely used, it can create a healthy and stimulating sound world for your family and profoundly enhance your child's growth.

How I Wonder What You Are

From the beginning of time, humankind has sensed the power of vibration, rhythm, and sound. Many cultures' creation myths describe a primordial sound or vibration that created matter from nothingness. The ancient Chinese and Egyptians considered music a fundamental element—one that reflected the principles governing the universe. It was believed that music had the power to uplift or degrade the psyche, to change the fate of entire civilizations. As a result, humans have made music throughout history to celebrate the passing of the seasons and mark passages in the lives of each member of the community, and have used rhythm to instill a sense of oneness among members of tribes and other groups.

Now, as one millennium ends and a new one begins, science is confirming the truth behind this age-old intuition. A recent article in *Science News* tells us that sound in the early universe, in the form of vibrational waves, may have helped orchestrate the striking pattern of galaxy clusters and huge voids we see in the sky today. We know that the moon itself vibrates, essentially "ringing" like a bell in a process known as spherical harmonics, probably in response to a long-ago meteor strike. In a similar fashion, tsunamis are created by the vibrational effects of earthquakes, which cause very small (yet detectable) waves that can grow enormously high. Music is simply a special case of this kind of vibration—a wave of energy that transfers some its power to us.

Music, the rhythm, tone, and vibration of sound, serves to *organize matter*—to create structure in space and time. Its effects are clear and measurable, not only on physical objects, but on biological entities as well. Reaching the human brain via the ear, music interacts on an organic level with a variety of neural structures. In fact, scientific research now indicates that this interaction has left its mark over the millennia on human physiology. The fact that a full two-thirds of the cilia in the inner ear—the thousands of tiny hairs that lie on a flat plane like piano keys—resonate only at the higher "musical" frequencies (3,000 to 20,000 hertz) suggests that at one time human beings communicated primarily with song or tone. One hypothesis is that human communication evolved from singing to primatelike grunts, before finally arriving at what we recognize as modern speech.

Perhaps this is why infants, newborns, and even fetuses universally display a remarkably high receptivity to music. Research has shown that a baby's brain arrives fully capable of recognizing such building blocks of music as key, pitch, and tempo. The systems that the brain uses to process music are either identical to or fundamentally entwined with the systems used in perception, memory, and language. Dr. Jamshed Bharucha, a psychologist at Dartmouth College, has suggested that the creation of organized music is an inevitable consequence of brain growth. When a computer model of the brain was exposed to music, the layer of cells responsible for recognizing individual notes quickly signaled another layer whose cells learned to recognize chords. Those cells in turn signaled a third layer, which soon learned to recognize groups of chords as belonging to different keys. Because Bharucha had not programmed the computer model to respond in this way, he theorizes that the patterns of music simply mirror the organizing structure of the developing human brain. Perhaps, he concludes, this is why we find music so pleasing.

"WITH MOZART, WE BECOME WHAT WE ARE."

My own formal introduction to the Mozart Effect began with a meeting in the early 1980s with the famed Parisian physician, psychologist, and educator Dr. Alfred Tomatis. The son of a musician, Dr. Tomatis became a specialist in hearing disorders, particularly those that affected professional instrumentalists and singers. Early in his career, he defined what came to be known as the Tomatis Effect—that is, the fact that the voice can only reproduce what the ear can hear. Having noticed that a group of French factory workers he was treating for environmentally caused hearing problems also had difficulty with speech, Tomatis soon realized that his opera singers' voice troubles were also often caused by hearing, or, more accurately, listening, problems. His success in improving his patients' expression by helping them listen better soon won him a wide following in the educational and musical community.

Dr. Tomatis continued his study of listening and its fascinating relationship to a wide variety of skills—including balance, posture, musicality, attentiveness, language ability, and expressiveness. "Man moves and stands upright because of the ear," he wrote. "Likewise, because of the ear he is able to express himself, listen, and think." Tomatis's second breakthrough came when he focused on the proximity of the hearing and emotion centers in the brain, and discovered that hearing disorders are often a reflection of emotional difficulties, and vice versa. To treat one effectively, he concluded, it was necessary to treat the other.

Following this trail of inquiry, Tomatis began working with children with psychological and learning disabilities, as well as children and adults with severe head injuries. Treating their disabilities via their hearing, he learned that different frequencies and rhythms of sound had remarkably different effects on his patients' state of being. High-frequency stimulation tended to reap the best results, increasing energy levels and creating a feeling of calm, while low-frequency sound often proved disorienting.

A third breakthrough in the 1960s—the recognition of the highly efficient effects of Mozart's music in particular—came when Dr. Tomatis combined what he had learned about the physical effects of sound with what he had learned from his study of embryology. He knew that the ear is the fetus's first organ to hook up to the brain's developing neural systems, and that the fetus begins to hear by the second trimester in the womb. Understanding that the mother's voice must serve as a kind of alternative umbilical cord for her developing baby, a prime source of environmental nurturing, Tomatis theorized that interference with hearing in utero and in the first years of life could lead to listening, learning, and emotional disabilities later on.

This would explain, Tomatis realized, why using high frequencies (such as those heard in the womb and, after birth, from mothers using baby talk) has such a helpful effect on the emotionally and developmentally challenged. With this in mind, he set about experimenting with all kinds of high-frequency sound for his young patients—as he put it, "all that could be acoustically recorded." His patients listened on headphones to both noise and music—classical, modern, traditional, and contemporary. He worked with music from Asia, India, and Africa. After all his experiments were completed and all data compiled, it turned out that two sound experiences were far and away the most effective in the children's treatment: the voice of the child's mother, filtered to omit all but the high frequencies the child once heard in the womb, and the music of Mozart.

My meeting with Dr. Tomatis (the first among many) confirmed my intuitive sense that Mozart's music produces an energizing effect quite different from that of other composers. In his two most important books, *Neuf Mois au Paradis* (*Nine Months of Paradise*) and *La Nuit Uterine* (*The Uterine Night*), as well as in my conversations with him throughout the 1980s and 1990s, this brilliant researcher has produced proof after proof that Mozart remains virtually alone on the list of very effective tools for treating a wide array of hearing and vocal deficiencies, as well as the disabilities to which they are related. "Whether in France, America, Germany, Alaska, Amazonia, or among the Bantus," he reported, "Mozart's music indisputably achieves the best results."

Obviously, there are other musicians of value, such as Handel, Haydn, Beethoven, and many more. But Mozart's music has an impact far greater than Bach's. "Exception among exceptions," Dr. Tomatis wrote, "Mozart has a liberating, curative, even healing power. With him, we become what we are."

Why Mozart?

Still, the question remains: *why* does Mozart's music work better than that of other composers? Is it simply that listeners prefer to listen to it? Or does it possess unique properties, eliciting universal responses that only now are yielding to measurement? Clearly, the rhythms, melodies, and high frequencies of Mozart's music stimulate and charge the creative and motivational regions of the brain. But perhaps the key to Mozart's greatness is that it all sounds so pure and simple. Mozart doesn't weave a dazzling tapestry like that great mathematical genius Bach. He doesn't raise tidal waves of emotions like the epically tortured Beethoven. His work doesn't have the stark plainness of a Gregorian chant, and it doesn't soothe the body like a lullaby. Yet his music is at once deeply mysterious and accessible and, above all, without guile. It is almost as though he was able to distill the beauty and order of the sound stimulation he experienced within the womb, and express it in a way that touches us on an equally essential level. Certainly, the wit, charm, and simplicity of his compositions allow us to locate a profound joy, and a deeper wisdom, in ourselves.

It is this ability to bring out the best in us that makes Mozart's music such a valuable aid in raising a child. The young, impetuous Wolfgang's life was far from ideal. He spent much of his childhood being shuttled from city to city in the back of a carriage, and he certainly had his imperfections as a person. Yet he embodies, in his music and in his own legendary exuberance, the vividness and personal brilliance we all hope to achieve at least now and then. For this reason, he and his music serve as a potent reminder of what many kinds of sound, from a mother's crooning to a Sousa march, can add to the life of a growing child.

Music for the Mind

If brain development is a process of incorporating patterns into increasingly complex systems, then music is a remarkably effective tool for providing those patterns. This patterning begins on a neuronal level in the womb and, after birth, continues with the patterning of movement, cognition, and the first experiences of social interaction. As the child learns to use words, the patterns of language and speech become tools for directing behavior and communicating. As words attain greater meaning, language patterns can be carried inward and organized into thinking and reasoning skills.

Practically every day, science provides us with visible evidence of the fact that music literally changes children's brains. New brain imaging technologies, such as magnetic resonance imaging (MRI) and the positron emission tomography (PET scan), have given neuroscientists a far more accurate view of the working brain than was previously possible. The PET scan, for instance, allows researchers to see which parts of the brain come alive during particular activities, and which parts are affected by different types of stimulation. Thus scientists can actually observe the brain in the process of developing. Having reviewed this new evidence, experts now agree that, though a child's inherited "nature" or temperament does play a part in his destiny, his environment is at least as important and perhaps more so. Throughout the early years of your child's life, beginning even before birth, his brain is irrevocably affected by his surroundings. What he sees, hears, touches, and otherwise experiences during this time influences not just his general development but actually affects, on a moment-by-moment basis, how his brain is wired.

At birth, a child's brain is in a remarkably unfinished state. Most of its hundred billion neurons are not yet connected into networks. His key task in early childhood is to seek out the interactions that will form and reinforce those connections. As the child begins to form attachments to his parents, family members, and other caregivers, and to explore his world, junctions, called *synapses*, are created by the thousands. By the time he is ten years old, his brain will have formed trillions. Each single neuron in his brain may be connected to as many as fifteen thousand other neurons, forming a network of pathways that is breathtakingly complex. If these synapses are used repeatedly in the child's day-to-day life, they are reinforced and become part of his brain's permanent circuitry. If they are not used repeatedly, or often enough, they are eliminated.

A high percentage of environmental input is provided through the ears, and the evidence is clear that from approximately eighteen weeks of gestation on, music plays a crucial role in the process of wiring a young child's brain. As your child is born and progresses through the years, music will enhance his physiology, his intelligence, and his behavior. Such effects are *real* and *measurable.* Studies have shown, for example, that:

- Music can calm or stimulate the movement and heart rate of a baby in the womb.
- Premature infants who listen to classical music in their intensive care units gain more weight, leave the hospital earlier, and have a better chance of survival.
- Young children who receive regular music training demonstrate better motor skills, math ability, and reading performance than those who don't.
- High school students who sing or play an instrument score up to 52 points higher on SAT tests than those who do not.
- College students who listen to ten minutes of Mozart's Sonata for Two Pianos in D Major (K. 448) tend to score higher on the spatial-temporal portion of IQ tests immediately afterward.
- Adult musicians' brains generally exhibit more EEG (brainwave) coherence than those of nonmusicians—and even differ anatomically in cases when the musicians began their training before age seven.

NATURAL LITTLE MUSICIANS

From the very beginning, you can make use of happy nursery songs, rhythmic rocking, lively dances bounced on the knee, and quiet sessions with classical recordings to bring harmony, mental stimulation, and joy into your baby's life. Western classical music, along with the chants, nursery rhymes, and songs of early childhood, contains all the essential rhythms and patterns of language, whether that language is English or Swahili. Thus, teaching your infant to appreciate music helps prepare his brain for mastering language's complex structure.

Once your little one begins to walk, music is like a steady hand to help his mind and body move together. As he internalizes a sense of rhythm that will eventually help regulate both his physical activity and the give-and-take of his social interaction, so he will learn to rely on familiar melodies and songs to create a daily rhythm that can form the bedrock of a secure, self-confident life. Invent songs based on his experiences, and you can strengthen your child's awareness of his environment, bolster his sense of personal identity, and encourage articulate speech. Meanwhile, highly organized music, such as Mozart's, can strengthen the connections among neurons that are also used in spatial temporal tasks, paving the way for later success in such abstract disciplines as higher mathematics and science even as it gives both you and your child pleasure.

As children grow, they begin jumping, running, and moving to music in a wide variety of ways. Do you remember how you moved your body to act out "The Itsy-Bitsy Spider," or to play "One Potato, Two Potato"? The opportunity in early childhood to sing well-known movement songs, dance to the beat, and make up musical stories helps children learn the way they learn best, through physical "hands-on" and "ears-on" experiences. Whether your child is performing a challenging violin exercise or just tapping his foot to the rhythm of a popular tune, his ability to move, think, and feel in a joyous, physical, creative environment is the essence of the Mozart Effect.

It is not necessary to be a professional musician or even to sing on key all the time to introduce music into your child's life. How well you pick out a tune on the piano or how gracefully you dance doesn't matter nearly as much as the passion and joy with which you share the world of sound with your child. Young children learn most effectively from those who love them, not those who display the most technical skill. As one mother, Sarah Cakebread, told me, "I will be the first to admit that I *absolutely cannot sing*, but it makes no difference. I sing all the time anyway. My children may tease me and my husband may shrug when I break out in song for no apparent reason, but it always makes them smile and puts them at ease. The songs we sing, from show tunes to Christmas carols or the Beatles, create a true sense of joy and even peace in our household. Personally, I believe there is a song for almost every situation, and if I can't find one, I simply put some music to my words."

So give it a try. It may not be easy at first to recall the songs of childhood or to move to the beat, but it gets easier with practice. Remember that hearing and making music is an enriching and positive part of human experience, and it can change children's lives. Make a connection with your little one through sound, and inspire and love him as the budding musician he naturally is.

CHAPTER 2

MOZART LISTENED TO MOZART

First Melodies of Life (Pre-Birth Through Birth)

Music is a swift weaver of deep feelings.
—Andrés Segovia

In the beginning, there was rhythm. The steady pulse of blood moving through the mother's body, the ebb and flow of her breath, the belly's deep, bass rumbles, the liquid movements inside the womb, and in the foreground, nearly drowning out the rest, the relentless gallop of the maternal heartbeat.

Time passed, in this world where the concept of time did not yet exist. Then, one day, a new sound emerged: the sharp, high-pitched trill of a woman's laughter. Soon, a second voice joined in, this one lower and more distant but exerting a powerful pull all the same. Then, who knows how many minutes, hours, or days later, a single, exquisite, musical vibration pierced the wall of the uterus. The sound was made by a violin, and it created an electrifying vibration, a sense of something completely new. A flood of other sounds followed the first—some painfully loud, some sublime, some nearly inaudible—each one a link in a golden chain that would lead the now actively listening creature out into the world.

The baby growing inside this womb would emerge in 1756 to become Johannes Chrysostomus Wolfgangus Theophilus Mozart—a child prodigy whose unique genius created some of the most inspiring music ever written. By the time he was four years old, this boy, nicknamed Amadeus (the French version of Theophilus, meaning "loved by God"), had already begun to compose and perform for his father's friends in the royal court. By age eight, he was writing great music, and in his short life of nearly thirty-six years, he created more than six hundred major compositions, including operas, symphonies, concertos, and great works for choir. From the beginning, Mozart's music was remarkable for its clarity and efficiency; its expressive, yet not overly emotional, content; and the subtle yet powerful ways it seemed to affect both the senses and the brain simultaneously. This composer's ability to evoke life's elemental joy through musical notes and rhythms was perhaps sparked during the months of his mother's happy pregnancy, when the playful sounds of his father's violin first reached his developing ears.

Surely no father ever harbored higher hopes for a child than Papa Leopold. The violin teacher, court composer, and future vice-conductor at the prince-archbishop's court in Salzburg, Leopold had already endured the deaths of five infants when his thirty-six-year-old wife, Anna Maria, became pregnant with Wolfgang. An involved, passionate (some would even say overbearing) parent and a consummate teacher, Leopold was determined that this pregnancy would produce the brilliant musical heir he felt he deserved. Unable to wait until his son exited the womb before beginning to instruct him, Papa Mozart created the *Toy Symphony* (originally attributed to Haydn) while his wife was pregnant. Its whimsical "little cuckoo," horn, and glockenspiel must have delighted the developing child and piqued his curiosity about the outside world. Wolfgang's four-year-old sister, Maria Anna, "Nannerl," did her part to welcome her brother, plunking and plinking on the keyboard that would become her lifelong companion. The Mozart family's lives had always been awash in music. It is impossible to believe, from what we now know about the effects of music on the developing brain, that the sharp, clear, pleasingly complex melodies they created failed to play a part in encoding young Wolfgang's nervous system with the transcendent, universal patterns and rhythms of nature.

Of course, the Mozarts did not know, as we do today, that certain high-pitched tones are the most likely to pierce the rhythmic din surrounding the fetus in utero to affect the development of the brain. They could hardly have suspected that the music of their own time would someday be considered a particularly effective tool for helping to create new neural connections, while at the same time bringing music to life in the developing baby's heart. Yet they must have sensed, as mothers and fathers always have, that their unborn child could somehow take in information, respond emotionally to stimuli, and even learn while still in the womb. Certainly, this intuition was at work in the ancient cultures of China and Japan, when a child's age was considered to be one year at birth. The centuries-old Asian practice of Tae-gyo focuses on educating the developing baby by exposing pregnant mothers to music and other artistic pursuits. Mozart himself shared this intuition so strongly that when his wife, Constanze, was pregnant with their first child, he reportedly composed his String Quartet in D Minor (K. 421/417) for the delivery. It is a valid intuition on which you, too, can rely to help your own developing child fulfill her greatest destiny.

THE DO RE MI'S OF FETAL DEVELOPMENT

Violinist Joanne Bath was completing her master of music degree at the University of Michigan when she learned she was pregnant with daughter Pamela. The news was most welcome, and Joanne still recalls the joy with which she combined her preparation for her final violin recital with her preparation for childbirth. The day of the recital arrived during the seventh month of Joanne's pregnancy. It was a celebratory experience, as she and her musician husband, Charles, performed, among other works, sonatas by César Franck and Aaron Copland. Many years later, Pamela followed in her mother's footsteps, performing her own master's degree recital on the violin. "She chose two of the pieces her daddy and I had played in my recital: the ones by Franck and Copland," Joanne told me. "She said they seemed so easy to learn, as if she instinctively knew them. She did not realize that she had heard them so much before birth."

Joanne isn't the first woman to sense that her unborn child must have been alert to what was happening outside the womb. Thanks to a great deal of recent scientific research, we now know unequivocally that the unborn child is indeed sensing and listening—first through the body, and later through the ear as well. Though the ear begins to develop only a few weeks after conception, it takes another four months or so for its connections to the brain to form sufficiently for the baby to begin to hear. But even before that time, sound's subtle vibration, moving through the skin and bones and filtering through the mother's body, becomes one of the first avenues of communication to the fetus. Simply put, your baby "feels" sound—including and especially the sound of your voice—for months before she ever hears it.

TUNE IN, TUNE UP
THE SONG OF LIFE

Over the past few decades, scientists have worked tirelessly to test the hypotheses underlying the notion that the fetus can hear, respond to, and learn from sound. As a result of their efforts, we now know that:

- The ear is the first sense organ to develop in the womb.
- The auditory system becomes functional three to four months before birth.
- After twenty-eight to thirty weeks of gestation, fetuses respond reliably to external sounds through changes in heart rate and behavior.
- Exposure to particular sounds can affect the auditory system in structural and functional ways.
- Familiarization with specific sounds before birth may induce a special sensitivity to, recognition of, and even preference for those sounds after birth.
- The human fetus is therefore capable of learning before birth at a level that can affect her postbirth behavior.

The auditory nerve, which transmits information from the ear to the brain, is the first sensorial nerve of the body to become functional. It establishes contact with all the muscles of the body through the baby's brain stem, working via the vestibular system, which regulates muscle movements and creates a sense of balance. In this way, the ear has a tremendous effect on the physical development of the body—which in turn affects your child's equilibrium and flexibility of movement.

By about the fifth month, connections in the baby's auditory system are mature enough to enable the brain to fully process sound. From this point on, your little one is eavesdropping virtually all the time. Your voice, having passed through the skin, muscle, and fluids of your body to reach her ears, probably sounds most often like sharp, high-pitched chirps. Still, the *melody* and *rhythm* of your speech (and of all sounds) arrive unchanged. According to Dr. Norman Weinberger, editor of the invaluable *Music and Science Information Computer Archive* at the University of California at Irvine, studies have demonstrated that Beethoven's Fifth Symphony reaches the fetal ear in a clear, identifiable state, even if some of the lower frequencies are missing. As your baby's brain continues to develop, her ear will act as a kind of tuning device, perceiving rhythm and pattern in the organized sounds that reach it and creating neural connections that mirror them.

As every expectant mother knows, developing babies don't just listen to sounds outside the womb—they also respond. (A powerful kick at a rock concert or loud symphony performance is a familiar experience.) Scientists have demonstrated that very loud sounds produce increases in heart rate, often with a startle response. Babies in utero have even been known to cover their ears in response to loud noise. New sounds tend to cause the baby's heart rate to slow down briefly, as if part of a cautious, 'What is it?' response.

But are these responses simple reflex actions, or do they demonstrate that your baby is actually "listening" and even thinking about what she hears? The evidence indicates the latter. From the time she can hear, if not before, your little one is not only registering, but also learning from the sounds that reach her. The simplest form of learning is *habituation*, which occurs when repeated information is no longer attended to or becomes boring. The listener has learned not to pay attention to that information. Infants demonstrate habituation when they first respond to a new sound (by, say, sucking faster on a bottle), stop responding as they grow used to it, and then respond again when a second new sound is introduced. Studies have shown that during the last trimester of pregnancy, a developing fetus behaves in just the same way—habituating to a sound that is repeated frequently, but responding through movement when the stimulus is changed.

More complex learning—for instance, *associating* one event with another—also takes place in the womb. In one study, mothers' abdomens were very gently vibrated in a way that did not produce a response in their babies. This massage was followed by a loud sound that did cause them to move. After pairing the gentle vibration with the loud noise several times, the babies began responding to the vibration alone, showing that they had learned to associate it with the loud sound.

Finally, the unborn child's ability to take in information and *remember* it later, even if not consciously, has been supported by a host of anecdotal and scientific evidence. Joanne's prenatal music instruction is one example of this intriguing process. Many studies have demonstrated that newborns clearly recognize and prefer music that their mothers listened to or sang during their pregnancies. Babies can even develop *literary* preferences while in the womb. In an experiment conducted at the University of North Carolina, infants whose mothers had regularly read Dr. Seuss' *The Cat in the Hat* during the final trimester of their pregnancies clearly recognized and preferred that book after birth. Research has shown that newborns can tell the difference between their mothers' language and a foreign one, and that at only two days after birth they greatly prefer their mother tongue.

The fact that babies are already listening to, learning from, and remembering music and sound in the womb suggests to a number of researchers that prenatal exposure to music can be used to enhance their development, and perhaps even alleviate or minimize some developmental delays. Researcher M. J. LaFuente developed a pilot program for prenatal teaching, beginning at the twenty-eighth and thirtieth week of gestation. At that time, mothers began listening to tapes that LaFuente had created of the basic elements of music, progressing over the weeks from a three-note major chord through more complex chords, for a total of fifty to ninety hours in all. In the months after birth, these babies developed many positive behaviors—including babbling, visual tracking, eye-hand coordination, exploring objects with the mouth, imitation of facial expressions, general motor coordination, and the ability to hold the bottle with both hands—significantly sooner than did a similar group of infants who did not listen to the music.

Certainly, some of these recent findings about the unborn child's ability to listen and learn merely echo common sense. After all, if infants born two months premature emerge with the ability to hear and process information, it seems logical to assume that they listened to and processed the sound that reached them inside the womb. Infants who are capable of feeling pain, joy, revulsion, anger, and love did not magically acquire these abilities in the birth canal. Children's brains develop from a few cells to a complex structure of billions of cells and trillions of connections—and that development hinges on a complex interplay between the genes they are born with and the experiences they have. This interplay between your baby's brain and her environment simply begins earlier than experts once thought.

In short, the sounds you make, the music you play, the words you speak while you are pregnant, can all send messages of love and encouragement to your unborn child—informing her about life outside the womb, and preparing her for her birth. As we will see, music that relaxes you will comfort and nurture your little one in direct, physiological ways from very early on; music that makes you happy will help you grow closer to your child; and highly structured classical music such as Mozart's will literally affect the architecture of her brain.

If you find it difficult to imagine communicating with your baby before she is born, engaging in the following playful exercise might ease the way. Sometimes, one of the most challenging aspects of pregnancy can be coming up with the perfect name for your baby. If you have tried one name after another, but have not yet found the one that clicks, try *singing* each potential name to a familiar melody. The following lyrics, for example, can be sung to the tune of "Twinkle Twinkle, Little Star":

Wake up, wake up little Jo,
Morning's come, it's time to go.
Wake up hands, wake up feet,
Wake up from your calming sleep,
Wake up, wake up, little Jo,
Now it's time for us to go.

Does the name sound right? If not, try another. Be sure to wait for a response from your listening baby, too. She is certainly listening.

The Primordial Sea: The First Four Months of Fetal Life

During World War II, music educator Grace Nash was consigned to a Japanese prison camp. Pregnant with her third child, she paced the narrow pathways of the compound, humming familiar songs to stave off fear and despair. During nighttime blackouts, she often played her violin for the other prisoners. And after her son was born, while she was still a prisoner, she sang to him constantly, especially while nursing, to calm her own fears and reassure him. To Nash's amazement, her infant son began to *sing* small phrases, syllables, and words long before he began to speak at the age of one. Nash later learned that this is a natural occurrence in cultures that retain singing as an integral and almost constant facet of daily life. From before birth, her son had learned to rely on music, and it remained his companion through his childhood. "On through high school, his singing or whistling indicated it had been a good day," she tells us. "Otherwise, oh dear . . ."

It is a terrible thing to experience extreme stress while carrying a baby. Yet rather than allow her natural fear and anxiety to damage her child, Nash used music to *protect* him by wrapping him in a warm cocoon of maternal melody and healing vibration. She had no way of knowing that scientists would one day discover that such activity can literally change a child's physiological development, altering his emotional perceptions, his level of volatility; in short, his character. The reason for this is that music and sound affect, for better or worse, a pregnant woman's mind, emotions, and physiological state. Her general level of stress, excitement, anxiety, or other emotion not only dictate her heart rate, quality of breathing, posture, and other physical aspects that in turn affect the baby within her womb, but they also lead to the secretion of hormones that pass through the placenta and into the baby's bloodstream. If the mother's hormone mix usually reflects a happy, loving, and relatively relaxed state of mind, that mix is fed directly to her baby. If her hormones very often reflect a state of fear or despair, her baby incorporates that message, too. In this way, a baby's most basic body chemistry is shaped, little by little, by her mother's emotions.

Since the 1950s, dozens of studies have proven that frequent or long-lasting emotional states in the mother can lead to organic changes in the unborn baby. One such study, conducted in Finland, focused on children whose fathers had died while the subjects were in utero or just after they were born. The rates of psychiatric disorders were substantially higher among the subjects whose fathers had died before, rather than just after, they were born—a clear result, the researchers concluded, of their mothers' stress levels during pregnancy. Their findings were seconded by another study that reported more underweight and frequently crying babies born to mothers who were very anxious during pregnancy or had a negative attitude toward motherhood.

Most scientific studies deal with the negative effects of a pregnant woman's experience, since researchers hope to develop ways to alleviate these effects. They also tend to focus on extreme examples of stress because the results are usually clearer and more dramatic. The normal, minor stresses of pregnancy (such as anxiety during an amniocentesis) also elicit a response (lots of movement from your baby), but this kind of fleeting anxiety is not likely to permanently alter your baby's brain. Probably a developing baby's experience of a passing "stress bath" is similar to a moment of intense fear for a very young child—unpleasant, but hardly life altering. Still, as you struggle to balance the everyday pressures of your own life against your natural desire for your baby to learn good things about this world, remember that in the timeless, largely featureless primordial sea that is the fetal environment, any major, repeated hormonal change is likely to loom very large. For your baby's sake, as well as your own, do what you can to relax as much as possible and keep your stress level down.

Spotlight on the Specialist

DR. THOMAS VERNY

In the nearly two decades since the publication of his landmark book, *The Secret Life of the Unborn Child*, Dr. Thomas Verny has continued to explore—through subsequent books and a professional journal he created that focus on these issues—the ways in which the pregnant woman's inner and outer worlds affect her baby's development. Dr. Verny is currently intrigued by new data demonstrating that the environment doesn't just interact with, but actually helps *create*, a baby's genetic heritage. We all inherit genetic potential from our parents. But just because a gene is inherited doesn't mean it will become manifest in our development. Recent research in cell biology reveals that environmental signals, particularly the mix of hormones in the mother's blood, are responsible for determining which genes are actually expressed by the baby. This means that the mother's emotions have an even deeper impact on her baby's development than previously thought.

As the fetus grows, the path of its development depends on the information received via the mother's blood, explains Dr. Bruce Lipton, one of the primary researchers in this area. If the mother's hormone mix frequently signals high anxiety or fear, the fetus selects from among its protective genetic programs—usually at the expense of growth. If, on the other hand, the mother's hormones relay the existence of a supportive, loving environment, then growth is encouraged. This process ensures with striking efficiency that the baby will adapt successfully to the world it will soon enter, and thereby survive.

To help parents turn this natural process to their babies' advantage, Dr. Verny recently created a collection of recorded classical music entitled *Love Chords*, accompanied by exercises aimed at providing stimulation for the baby, encouraging healthy development and enhancing the relationship between mother and child. "For a long time other scientists believed that genetics was destiny," Verny says. "But we know now that brain growth and organizations are continuously responding to the environment. Music is part of that."

It's difficult to accept at first that something we tend to consider a simple pleasure—the transcendent compositions of the classical composers, the soothing lyrics and melodies of good popular music, and even the simple, rhythmic melodies of children's songs—can actually affect who a child becomes. Even more surprising, though, is the notion that music can have perhaps its most powerful influence via the mother's bloodstream, before the unborn baby's hearing apparatus has developed. Still, the evidence is in—your growing baby is open to as much good hormonal news as you can give her. By consciously relaxing, stimulating, and expressing yourself through music, you will be able to counteract the fleeting anxieties and fears of a normal pregnancy, and create a supportive universe for your little one. In short, you can turn your pregnancy into a natural Head Start program for your child.

In order to best use music to help modulate your emotional state, it's important to experiment with a variety of musical selections to find those that seem to work best at retuning your mind and body. As always, Mozart is a good choice when we seek the attendant benefits of music. (Besides, the music is beautiful!) Find a quiet time to sit alone in a dimly lit room, and sample the Andante movement from Mozart's Piano Concerto No. 21 in C Major (K. 467); the Rondo from Sonata in F Major for Violin and Piano (K. 376), and the Rondo-Allegro from *Eine Kleine Nachtmusik* (K. 525). How does each of these selections make you feel? Do you sense that you and your baby are in tune while listening to Mozart together? Pay attention to the differences in your pulse, body temperature, state of alertness, and general mood that each selection creates.

Keep in mind that soothing music is not the only type of sound that's good for the two of you. Faster, allegro tempo pieces activate the most alert brain wave state—the beta state—allowing you to work, think, and exercise with optimal energy. Even some pop music that raises your agitation level a little may benefit your child. If used sparingly, such a mild form of "sonic caffeine" may help focus your baby's attention momentarily, and perhaps inspire a dance kick. As Dr. Verny points out, babies spend about 95 percent of their time in utero sleeping, so stimulating music can help them wake up and perceive sound. If you prefer light jazz or gospel to Mozart, include it once in awhile in your routine. It might also be fun to collect songs that simply make you happy—sentimental favorites, children's folk tunes that bring happy thoughts of motherhood to mind, or even lullabies from distant lands.

Once you have selected your favorite compositions and songs, record them on one or more tapes, and keep the tapes at hand—in the car, by the bed, in the living room—for those times when the stresses of life begin to bring you down. As you listen, close your eyes and visualize your growing baby also listening inside you. Slow your breathing, concentrate on inhaling deeply and exhaling fully, and send her a message from Mom that no matter how hard life can be now and then, the simple joy of living overcomes all challenges.

CELEBRATING LIFE WITH SONG

In the tiny French village of Pithiviers, obstetrician Michael Odent began organizing group meetings around a piano so that expectant mothers could sing together on a regular basis. Most expectant mothers need much more social and emotional support than is available through a monthly meeting with an obstetrician or midwife, he felt. He hoped that by creating a friendly and welcoming songfest for his patients, the women would form emotional attachments with the clinic and with one another.

In Valencia, Spain, midwife Rosario N. Rozada Montemurro helped organize a twice-a-week singing group for pregnant woman at the health center where she works. The group chooses from a repertoire of traditional lullabies in Spanish and the local dialect, so that the mothers will also be able to sing the songs later, when they are alone with their newborns. These deeply familiar songs embody the mothers' desires to cradle their as yet unborn babies, Montemurro writes, in a way that folk songs and lullabies of other cultures never could. Some of the pregnant women can remember their own mothers and grandmothers singing the songs to them as they were lulled to sleep to the rhythmic, monotonous *tac-tac* of a rocking chair against a wooden floor.

"As we sing, we allow our almost-hidden dreams to express themselves, we let our fears emerge, we encourage ourselves to feel and express the whole range of our emotions, from friendship, affection, to pain and loss," Montemurro explains. Not all the songs are relaxing; some are invigorating, joyful, and enlivening. The women sing in two-, three-, and four-part harmonies, clap, tap their feet, and pat their bellies to transmit the rhythms to their unborn babies. The pregnant women report that they not only enjoy the singing, but can also feel their unborn children participating with harmonious and spontaneous movements. The experience is so pleasurable that the singing groups often continue after birth as well, encouraging close contact among mothers and children for years to come. Fortunately for North American mothers-to-be, many early childhood music programs have also begun to provide group music-making sessions for expectant parents.

As these examples suggest, making music can be an even more powerful means of reaping the rewards of sound than passively listening to even the highest quality recordings. Both activities can affect your hormonal levels and influence your pulse rate and blood pressure. But singing also sends energizing vibrations through your muscles and bones, invigorating both you and your child; improves your breathing, thus rushing nourishing oxygen to your baby; and helps you actively rid yourself of anxiety and unhappiness. If you find singing an effective way to forget your cares at the end of the day, consider forming a singing group of your own. Certainly it's a wonderful way to break down inhibitions and create a group of supportive friends as you prepare for the transformation of childbirth. If you're more solitary by nature, there's nothing wrong with doing your singing in the shower—or at home with your baby's father, so that your child can grow to love his voice, too.

Toning—a related, soothing activity that involves making sound with an elongated vowel for an extended period—offers expectant mothers the added benefits of increased focus and serenity at the beginning or end of a busy day. The vibrations of toning can give you and your child a feeling of internal massage. It oxygenates the body, deepens breathing, and relaxes the muscles. A practice dating back hundreds of years, toning can also be used to open natural channels of energy in the body, just as we use electricity.

Anyone can learn to tone. In *The Mozart Effect*, I described a five-day "toning course" that takes only five minutes each day. By following this course, you can become a proficient toner before the week is done—and your child can benefit from your increased calm and sense of well-being.

On the first day, sit comfortably in a chair, close your eyes, and spend five minutes humming—not a melody, but a pitch that feels comfortable. Relax your jaw and feel the energy of the hum within your body. Bring the palms of your hands to your cheeks and notice how much vibration is occurring within your jaw. This five-minute massage will release stress and help you relax.

The next day, instead of humming, make an *ahhhhh* sound. The *ahhh* immediately evokes a relaxation response. You produce it naturally when you yawn, and it can help you both wake up and go to sleep. If you feel a great deal of stress and tension, take a few minutes to relax your jaw and make a quiet *ah*. There is no need to sing. Just allow the sound to move gently through your breath. After a minute or so, you will notice that your breaths are much longer and that you feel more relaxed.

On the third day, the toning sound is *eeeee*. This is the most stimulating of all vowel sounds, awakening the mind and body. When you feel drowsy while driving or are sluggish in the afternoon, three to five minutes of a rich, high *ee* sound will stimulate the brain, activate the body, and keep you alert.

The *oh* sound, used on the fourth day, is considered the richest of all by many people who tone or chant. Make the *oh* sound. If you put your hand on your head, cheek, and chest, you will notice that the *oh* vibrates most of the upper parts of the body. Five minutes of the *oh* can change the skin temperature, muscle tension, brain waves, and breath and heart rates. It is a great tool for an instant tune-up.

The fifth day is your time to experiment. Start at the lowest part of your voice and let it glide upward, like a very slow elevator. Make vowel sounds that are relaxing and that arise effortlessly from the jaw or throat. Allow the voice to resonate throughout the body. Now explore the ways in which you can massage parts of your skull, throat, and chest with long vowel sounds. Let your hands trace the upper parts of your body very slowly, and you will see which vowels emit the strongest, most stress-releasing energy.

Toning doesn't have to be a solitary activity. On nights when you're feeling tense or uncomfortable, invite your partner to "spoon" with you and your unborn child. Lie on a bed with your back to him, allowing him to nestle up behind you like two spoons in a drawer. Let him begin toning first. Feel the low rumble of his vocal vibrations move through your body—and your baby's. Now add your own tone. Visualize your child, bathed in the comforting vibrations of her parents, enveloped in audible love. When you've had enough of this position, roll over to face your partner. Tone again with him, feeling the baby roll, stretch, and (perhaps) sigh with pleasure between you. This is a real musical trio.

THE FETAL ACADEMY: FIVE MONTHS OF STIMULATION

"Dear Mr. Campbell," wrote Canadian music educator Dorothy Jones, "our son listened to Vivaldi's Violin Concerto in A Minor in utero and for several weeks after birth. At six weeks he was vocalizing every time he heard it, and his vocalizations were readily recognizable as the opening theme of the first movement. When this child began kindergarten, he came home from school in the first few days and announced that the bell at school was an A. We checked it out, and he was right. Further checking confirmed his perfect pitch. When he was young, teachers always mentioned his wonderful listening ability at school, as did his piano and violin teachers. It was this family experience that led me to develop the prenatal and baby program as a component of our Suzuki Centre."

Stories abound featuring babies and young children who spontaneously demonstrate knowledge they could only have developed before birth. The parent of an eighteen-month-old reports that her son began singing a particularly melodic piece from Beethoven's *Pastorale* upon waking in the morning. She had listened to that piece, she said, several times a day while she was pregnant. In Austin, Texas, a mother describes her daughter's remarkable and unexpected proficiency for math—a talent, the mother is convinced, that is attributable to her child having listened to Bach's cool, logical, symmetrical music throughout her prebirth life. A Baltimore dad wonders whether his son's talent for dance has anything to do with the mother's brief involvement with African-drum exercise classes when she was expecting.

It is fascinating to try to imagine the experience of first hearing sound—the low pounding of the mother's heartbeat, or a glittering fragment of her voice. We have all experienced this miracle, yet the moment lies buried deep in our psyche, beyond recall. Still there is no doubt that at about five months gestational age sound bursts the unborn baby's prenatal world wide open—propelling her out of her silent, unified sphere into a new state of individuality.

No longer as dependent on your hormonal messages for help in understanding the outside world, by the fifth month your baby can now perceive and process information for herself, through sound. From this point on, her exposure to an enriched aural environment such as high-quality music will stimulate the growth of dendrites—the branches that extend from the brain's neurons to form the connections that make up the neural web. When you think about it, this makes sense: music is a form of patterned vibration, and the brain stores and recalls information most efficiently in the form of patterns. The more that developing babies are exposed to appropriate music, the more chances music has to pattern brain growth in an organized, harmonious way. In fact, studies have demonstrated that one or two sessions of low-volume classical violin music a day for the third-trimester fetus can lead to advances in gross and fine motor activities, linguistic development, and certain cognitive behaviors that are clear both at birth and later.

Sound consists of a variety of building blocks that affect your unborn baby in different ways. Rhythm—experienced through your familiar heartbeat, the pleasant rocking sensation you create as you walk or rock in a rocking chair, and finally, through the rhythm of human speech and music—is the first musical quality that your developing baby apprehends. Rhythm is also one of the first concrete ways you can communicate with your child. What an amazing experience it is to find that your baby—perhaps even before she begins to hear—can communicate back!

At around the middle of the fourth month or the beginning of the fifth, start trying to initiate a conversation with your child. Begin by lightly tapping out a simple rhythm on your abdomen. You may have fun vocalizing the taps, as though they are short and long dashes in

Morse code: dah-dit dah-dit dit-dit dit-dah. Tap the rhythm out once, then twice, then a third time, pausing between each "message." Now wait a day. Then tap the message again. Does your baby's movement stop as soon as you begin tapping? That means she's listening. If you vary the rhythm you've been using, does she indicate that she's noticed the difference by suddenly moving a bit?

As the months progress and you continue your tapping game, you may find that your baby actually taps back or kicks back once in a while. If so, move your tapping to another place on your belly and see if she kicks that place next time. Some parents manage to get quite a duet going with their unborn children, tapping very short, simple rhythms that their babies then tap back!

Dr. F. Rene Van de Carr, a California obstetrician who, with partner Marc Lehrer, developed a "prenatal university" aimed at helping his patients communicate effectively with their unborn children, recommends conducting such tapping sessions at around the times when your baby is likely to be in a responsive, alert state. Usually, this is the case about half an hour to two hours after a meal (you will know she's awake by her kicking and other movements). Communicating with your baby at roughly the same times each day conditions her to play and learn during those times—a routine that you can continue into babyhood. It also helps you get used to the idea of considering your child's condition before imposing an activity on her. Thus, the give-and-take of the ideal family communication can begin before your baby is born.

At around five months, your baby's ears and brain are wide open to every bit of stimulus she receives. From now on, she will become increasingly adept at differentiating pitch in the voices, music, and random sounds that reach her world. This is important because pitch differentiation is the first step in acquiring language, as well as musical ability and, eventually, reading skills. If you haven't been talking, singing, or otherwise verbally communicating with your baby before now, it's time to begin. How you do this depends entirely upon your preferences. Some parents love to read or *sing* the bedtime stories they've already begun collecting for their babies. Mothers have the option of singing in the shower or talking to the baby as they do household chores. If you haven't already been doing so, start talking along with the tapping game described above. Name the action as you or your baby does it ("Tap, tap, tap . . . kick!"). Your baby will be listening, and your voice will act as positive reinforcement for communicating with her even more.

The fact is, nothing is as pleasingly stimulating to your baby's brand-new ears and growing neuronal network as your own voice—no matter what your friends and relatives think of it. Unlike other voices, yours is doubly amplified: it reaches your baby from *outside* your body, like any other sound, and from *inside*, as the waves of sound vibration pass through your bones and tissue and envelop her in a gentle massage. Until (and for at least several weeks after) birth, your voice will remain by far the most important and riveting sound she hears. During the second half of your pregnancy, your baby will study every aspect of your vocal tone and rhythmic syntax. As a result, she will emerge at birth "imprinted" with your native language. If you speak English, your Spanish-speaking neighbor's voice may not particularly interest her at first. If your mother, who speaks with a foreign accent, addresses her brand-new grandchild, your baby may even start to cry!

A MUSCAL RECIPE

A GIFT FOR THE ADOPTED CHILD

Knowing that children begin learning language before birth is important for adoptive parents as well as for those who feel they must give up their child. The decision to give up a child for adoption is never easy, Colorado health practitioner Willow Pearson reminds us. As all birth mothers who have given up children for adoption, and as all adopted children know, even in the best of circumstances there remain a host of complex emotions that go along with adoption. One of the reasons for the adoptive infant's grief or confusion may be the loss of any sound connection to her prenatal life.

"Imagine, then," continues Pearson, who has also studied music therapy, "what it would be like as an adopted child to hear a song that your birth mother had sung frequently and lovingly when she was carrying you—especially if it's in a language that you don't otherwise hear anymore. Imagine what it would be like as a biological or an adoptive mother to offer such a gift to your baby, in the process of separation.'' It was this idea that inspired Pearson to found the Birthsong Project in association with Boulder Community Hospital. Pearson suggests, in cases of open adoption or when the birth mother is known, that adoptive parents ask the mother to record the songs or music her unborn baby heard or seemed to like the most. Then, after the adoption, these same songs played for the infant or sung by the adoptive mother can act as a ''sound bridge'' connecting one life to the next. If the mother isn't known, recordings of traditional lullabies from the child's birth culture may have a similar effect.

UR-SONGS, YOUR SONGS

As your child grows to know and love her mother's voice, so will she learn to love the music you offer her. During the final trimester, appropriate melodies can further spur her brain development in exciting new directions. One simple song to begin with consists of the three-tone melody that accompanies "Ring around the Rosey," "It's Raining, It's Pouring," or the teasing chant "I've got a secret." This three-note ur-song is generally considered a universal melody by musical researchers, probably genetically encoded in the brain in the same way that songs are for birds. The same three notes, along with the "One, two, tie my shoe" rhythm, have been found among children from all cultures and backgrounds. By patting your belly in rhythm as you sweetly sing this song to whatever words you like ("Mama loves her baby" works very well), you are giving your child a number of brand-new and very intriguing sound concepts to ponder. Once you have awakened her ears with these universal sounds, it's easy to move on to other, more soothing children's songs. By doing so, you have taken the first steps toward encouraging your baby's future language comprehension and musicality.

During the final month or two of pregnancy, your baby is far enough along to benefit from more sophisticated sounds. At this point, it's a good idea to begin singing a wider variety of songs to her (anything from "She Loves You" by the Beatles to "America the Beautiful"!), and tell her stories using a few simple words that she might come to recognize. Some parents have even included opera music, symphony recordings, poetry, and religious readings in these final sessions. Give some thought to what you choose—your baby will probably insist on hearing it throughout the years to come!

A Sound Solution

SHE WANTS HER OWN STEREO?

Impressed with the evidence of music's beneficial effects on the unborn baby's brain, Indiana obstetrician David Min was inspired to create a gadget he calls the Rock-a-Bye Music System to help the process along. His invention is simply an apron with three adjustable pockets—one facing outside to hold a portable tape player, and two pockets facing the stomach to hold small speakers that are positioned over the mother's womb. By wearing it, you can play recordings of family members' voices, music, or rhythmic sounds for your baby while driving, washing dishes, or doing other chores. Some of Dr. Min's patients have found that playing music for their babies at bedtime helps them to calm down and be still. Others have reported feeling a "different" movement when the baby was listening to recordings of her parents' voices. During a difficult labor, you can use the tape player to soothe your baby with classical music, which may in turn decrease your baby's heart rate and increase the likelihood of a safe delivery. Keep in mind, though, that we do not know exactly *how much* prenatal sound is healthy. Beware of more than a half hour at a time, and keep the volume low.

Curious about whether prenatal exposure to music truly affects children's long-term development, researcher Donald Shetler initiated a study in which unborn babies listened to music daily in two five- to ten-minute sessions, through stereo headphones placed on the abdomen. After the babies were born, he noted that when he played the same music used in the prenatal sessions, the infants responded with instant movement, a fixed gaze at the source of the music, and in one case, reaching out to the sound source. Shetler also discovered more long-term and profound effects. Once the child had begun using spoken language, he observed an "early development of highly organized and remarkably articulate speech"—implying that prenatal music stimulation enhances verbal expression.

Music is a wonderful boon for developing babies, but it is possible to have too much of even this good thing. If five minutes of prenatal stimulation twice a day is good for your baby, that doesn't mean that three hours of Wagnerian opera is better. Though there has never to my knowledge been a proven case of overstimulation of an unborn baby, it stands to reason that the unborn child will respond to too much sound and movement by starting to turn in on herself and "tune out the noise" through the process of habituation. This tuning-out behavior could cause problems in language development, physical coordination, and even reading and writing, as the entire hearing system fails to respond to the corrective guidance of incoming sound.

Loud noise—especially loud, low-frequency sound—can also be a problem for your baby. This includes not only the frightening noises of shouting adults, but also the chronic din of certain work sites and public spaces. While you are pregnant, avoid noisy places as much as possible, as well as those with excessive vibration. United States law dictates that you should be able to request a transfer to a quieter work site in most situations without being penalized by your employer. For more information, contact the Occupational Safety and Health Administration (OSHA) at (800) 321-OSHA.

FIRST MOVEMENT: BONDING MUSICALLY WITH YOUR BABY

It is astonishing, really, to consider that for most of the past so-called modern century, babies were not considered conscious until after they were born. Any mother could have told these scientists differently—and you can, too, now that you have rocked, tapped to, and serenaded your baby and noted her eager response.

From at least the sixth or seventh month of pregnancy, mother and child are in a state of intense communication that ideally grows more lively as the date of delivery draws near. Your baby is growing adept at comprehending the emotional tenor of your voice as you read or talk to her or chat with others; the quality of your rhythms (whether rushed and erratic or deliberate and soothing); and even the faint physiological traces of your thoughts. If this last assertion seems far-fetched, consider the study by Dr. Michael Lieberman showing that during the third trimester, if pregnant women who smoked even thought of smoking, their babies' heart rates sped up—a sign of fetal agitation.

Imagine spending days entwined in bed with your partner, your ear pressed up against his chest. Wouldn't you know when he became frightened, fearful, excited, joyful, or relaxed? Of course you would—his heartbeat, the heat of his body, the tension of his muscles, the tone of his voice, and the rhythm of his movement would tell you everything you needed to know. This is how your unborn baby lives, except that for her these emotional responses are the only reality, not just Mom's passing moods.

Whereas in the first trimester your baby was affected on an organic level by your habitual emotional perceptions, she is now increasingly able, through her developing consciousness, to perceive more fleeting and finely tuned feelings, such as self-confidence, sudden joy, and pleased excitement, as well as, unfortunately, frustration, ambivalence, and sadness. This awareness in your child doesn't mean you have to tiptoe through the rest of your pregnancy, always on the lookout for any minor negative emotion that might affect her. It does mean that your baby is getting to know you better, and is, on some level, pondering what your emotions mean. It makes sense, then, to put your best foot forward, reassuring her that she's wanted and loved, that you can't wait to see her, that the world is ready to welcome her with open arms.

Mozart's music is a wonderful way to communicate positive messages to your baby. His melodies, to which the developing baby attaches itself with such relish, tell us that life offers a wealth of joy, serenity, and stimulation. Playing a recording of the *Variations on Ah! Vous dirai-je, Maman* (K. 265), the "Twinkle Twinkle, Little Star" variations, will put you both in a good mood even as its bright variations stimulate her brain.

For the developing child, music is a language all its own. It communicates, and through its complex patterns and rhythms it creates a bond between the family and the unborn child. The cheerful sounds of children's songs sung by the family introduce the baby to warmth of the singers' voices (no matter what their tonal quality), and tell her worlds about the emotional relationships between mother and father, parents and children.

In one community in Uganda, a child's birth is believed to occur even before conception, when the mother first "conceives" of her in her mind. The mother's thoughts lead to the creation of a ululating song that embodies her dream of the child. Once she has sung the song, the mother teaches it to her partner, and then sings it with him as they make love. Throughout her pregnancy, the woman continues to sing the song to her little one, and during birth the midwives and old women welcome her with what they believe is a now-familiar melody.

In similar ways, Willow Pearson of The Birthsong Project works with expectant parents in creating a unique birthsong for their babies. She encourages the parents to talk to her about their wishes for and early perceptions of their child. Then she reflects back to them what she has heard through music—creating lyrics and melody from their words. This prompts the parents to respond through musical play and improvisation, using just the voice and simple percussive instruments—such as a shaker or rattle—to create a heartbeat rhythm. Though few of the parents have any formal musical experience, they are thus able to craft a beautiful, simple lullaby that captures the essence of the love they feel for their baby.

You, too, can easily create a birthsong for your child—a song that will become familiar to her long before she is born and comfort her in the years that follow. Simply sit down and think about the feelings you have for this little person. Tap out a heartbeat rhythm with a shaker or just your hand, and relax until the words begin to flow. Pearson also recommends creating a "family song," singing the names of all the people who are waiting to welcome your baby into the world. You may be amazed to realize how many there are. This knowledge can reassure you as much as your child as you face the sometimes intimidating life transition of giving birth.

As the final days of your pregnancy approach at what sometimes seems a glacial pace, be sure to take advantage of the relative calm before childbirth. Find a quite place where you can sit with your baby and soak in a little pure silence. If possible, install a rocking chair so you and your child can relax into its soothing rhythm. Don't talk to your baby during these quiet times but gently stroke your belly to let her know you're thinking of her. If your little one has been kicking and moving a lot, you will enjoy feeling her slow down, perhaps roll over, and lazily stick a foot out to meet your hand. These silent moments alone together are some of the most precious of pregnancy. Enjoy them while you can.

TUNE IN, TUNE UP

SWEET NOTHINGS

No one's life is entirely stress free, and pregnant women generally have enough to worry about without wondering whether their unborn children are getting anxious, too. You can reassure your baby and yourself that no matter what challenges loom ahead in the short term, the long-term message is a positive one, by taking the time at the end of each day to "sing yourself down," giving both you and the baby a five-minute voice break. To do this, push the cares of the previous hours out of your mind, make yourself comfortable in a favorite chair, and practice inventing little nursery songs to amuse your child. Keep the melodies simple (perhaps borrowing the melodies of "You Are My Sunshine" or "Are You Sleeping?"—also known as "Frère Jacques"). Tap the rhythm lightly on your stomach as you sing. Practice varying the emphasis and pitch a bit as you play with the music. Remember, your baby doesn't care whether you have a professional-sounding voice. She just wants to know you're happy to harmonize with her.

A MUSICAL BIRTH

"I was alone with my baby all that long night, for fourteen hours of labor," Ellen, a single mother in Philadelphia, wrote to me recently. Though Ellen had been joined by friends in the birthing room for a few hours that evening, it was now late and everyone had gone home. "A very kind nurse checked in on me every once in a while. Besides that, there was nothing but the waves of contractions." Fortunately, Ellen had brought a tape player to the hospital with her, stocked with the Mozart and Vivaldi tapes she'd listened to throughout her pregnancy. "I put on the first tape—I think it was some violin concertos. And all of a sudden it was like Max, my baby, and I weren't alone. We had our music with us. I felt as though it brought us together in our struggle, in a way. Also, I found I was able to ride the contractions much more easily with the music playing.

"They didn't let me take the music with me into the operating room," Ellen added (she had to have a cesarean section). "But it was still playing in my head. I guess you could say Mozart was my birthing coach. He helped give me my beautiful little boy."

Traditionally, music has long been a part of the physical process of childbirth, from the ritual dances and songs prescribed in Nigeria to the belly dances of the Middle East—movements originally intended to assist women in bringing a child into the world. Today, women still use music to accompany their movements and improve their concentration during toning and childbirth-preparation exercises, and to soothe their anxieties over the approaching "big day." Certainly, music makes any exercise routine more fun, which means you may be more likely to do it at home. Recordings with distinct, even rhythms or songs with strong vocal melodies work best when you're moving to the beat, according to Janice C. Livingston, a childbirth educator in Ocala, Mississippi. For exercises done to counts of four or multiples of four, use music with a 4/4 tempo. A waltz (3/4) tempo works better for a three- or six-count exercise.

As always, though, listening to music is only half the fun. Making music can be an even more effective way of turning the challenge of childbirth into a labor of love. Mary, for example, a good friend who sings in a local church choir, was pregnant with her first full-term child at the age of forty-three. Due to her age and the fact that she had previously experienced several miscarriages, Mary's doctor had convinced her to have an elective cesarean. As it happened, another woman in the choir, forty-year-old Susan, was also pregnant and had agreed to a cesarean. Both women continued singing in the church choir up until their last month of pregnancy, attending rehearsals once a week as well. When the time came for their children to be born, both babies were delivered vaginally, and easily. Both Mary and Susan insisted that singing, particularly in concert with others, had made all the difference. "To sing, you have to learn to breathe correctly," Mary pointed out. "I spent my entire pregnancy breathing rhythmically, using my abdomen, expanding my thorax, increasing my lung capacity, and inhaling through my nose and exhaling through my mouth just like a good Lamaze student. For me, singing was a very fun and energizing form of physical therapy."

An increasing number of women can attest to the benefits of music while undergoing labor as well. Anecdotes abound about its power to make time pass more easily, to make contractions more bearable, to make the mother feel more actively involved in the birth process, and to bring families together as they await the arrival of a new member. Listening to music during the actual emergence of the baby inspires feelings of euphoria in many mothers. Listening to the same tape with your infant in the following months provides you both with a feeling of continuity and companionship, and helps you relax. In one study, more than half the women who used music during pregnancy and childbirth reported having "easy" babies who seldom cried.

Clearly, playing beautiful music during childbirth is good for babies, too. "Birth is the first prolonged emotional and physical shock the child undergoes, and he never quite forgets it," writes Dr. Thomas Verny in *The Secret Life of the Unborn Child*. "He experiences moments of incredible sensual pleasure—moments when every inch of his body is washed by warm maternal fluids and massaged by maternal muscles. These alternate with others of great pain and fear." What better way to reassure a newborn—to wrap him in a warm security blanket of familiarity and beauty—than by accompanying his entry into this world with recordings of the transcendently beautiful music he has listened to all his life?

Lamaze breathing techniques adapt easily to music, Livingston points out, since they are basically an exercise in rhythm. Throughout the latent phase of labor, when you will be breathing slowly through your chest, a 4/4 tempo with a distinct drumbeat makes sense. Once in active labor, when shallow chest breathing will help you cope with the increasing strength of your contractions, it's better to play music with a faster tempo. You might also want to turn the volume up to enhance your concentration.

The selection you choose for this final birth phase is, of course, an entirely personal matter. In the years that I have been researching music and birth I have been truly amazed to learn what works for birthing mothers—everything from the soundtrack of *2001: A Space Odyssey* to rock music, Beethoven, Dvorak's *Serenades*, and even the charming songs of Peter, Paul and Mary. Whether it lasts for one hour or twenty-four hours, birth is a potent, rhythmic process that passes through many phases, just as there are many movements in a symphony. Music can bring a continuity to this process. Have your husband act as disk jockey. When it's time to push, let rhythmic music assist you, and when it's time to rest, let the kind, inspirational music that you love soothe and inspire you.

To answer the increasing demand for greater access to music during childbirth, some hospitals are setting up sound systems in their delivery and recovery rooms. Dr. Fred Schwartz, an anesthesiologist in Atlanta who has been a pioneer in this area, has used music in his practice for over twenty years and is convinced of its ability to decrease the mother's stress response, which can be good for a baby in distress as well. In fact, in a study completed by a pair of therapists in Texas, pregnant women who listened to music they liked during labor were only half as likely to need traditional anesthesia during the birth.

Again, *making* music can benefit you even more than listening to it, in childbirth as well as during pregnancy. If you took the time to explore the energizing effects of tone in earlier months, you can well imagine how it might invigorate you and your child throughout the birthing process. Visionary childbirth educator Beverly Pierce, who has worked extensively with toning in labor and childbirth (as well as during pregnancy), believes that toning also helps women focus and cope with pain. When she asks her students why they toned during childbirth, they surprise her with how often they claim that toning gave them a sense of personal power. "It connected me with nature," one woman told her. "I felt like I had more control. The more it hurt, the lower I'd bring the sound."

Obstetrician Michel Odent has pointed out that as a woman moves more deeply into labor, her awareness shifts from the outer world to her own inner world. If active labor is associated with lower brain wave frequencies, then by slowing and lengthening your breath through toning, you may be able to alter your consciousness and thus enhance the childbirth experience.

Whether you bring your baby's favorite cassette tapes into the delivery room (the ones you've played for her for the past six months), tone right up until the critical moment, or belt out a song as your baby emerges from one world to another, don't let this moment go uncelebrated by music's natural power and grace. A symphony, a grandmother's favorite lullaby, a national anthem, Dvořák's *Serenades*, or Mozart's ethereal *Magic Flute*—the selection you pick to mark this momentous occasion is less important than being able to hold your child in your arms someday and say, "This is the song that led you into the world."

A Mozart Musical Menu*

- *Variations on Ah! Vous dirai-je, Maman* (K. 265). Whether they start you singing "Twinkle Twinkle, Little Star," "Baa, Baa Black Sheep," or the French folk song on which they were based, these sparkling variations will stimulate your growing baby's brain development and cheer you both up at the same time.
- Andantino from Flute Quartet in C Major (K. 171) (258b). In this playful excerpt from the flute quartet, Mozart almost tells a story. The light, active and fresh music is easy on the ears, excellent "ear candy" for your developing baby.
- Andante from Symphony No. 25 in G Minor (K. 183). In this symphony, Mozart uses a perfect "Go to Sleep" theme. Listen carefully and you will hear the music saying "Go to Sleep" in its melody. On nights when your baby refuses to settle down, try playing this for her as you sing or chant "Go to Sleep" every time you hear the theme.
- *Love Chords*. Dr. Thomas Verny's compiled collection of soothing and stimulating musical selections and exercises for the mother-to-be and her child.

*All "Musical Menu" suggestions are available on my specially prepared CDs for infants and children. Please refer to "The Music of Mozart" at the end of this book for details.

CHAPTER 3

Cry Baby, Lullaby, and Itsy-Bitsy Songs

The Gift of Music (Birth to Six Months)

Like music on the waters
Is thy sweet voice to me.
—Lord Byron

How different the world is outside the womb. Where once your baby felt the warm, supportive pressure of the amniotic sea, now there is nothing but empty air. Where once faint, high-pitched melodies filtered softly through the layers of your body, now honking horns, ringing telephones, and blaring TVs assault his delicate ears. A faint glow may have occasionally pierced the uterine wall; now brilliant light tests the tolerance of newborn eyes. Faced with a bewildering avalanche of sensory input, your baby searches actively for any sign of familiarity—for the predictable patterns of sound and movement that helped form him. The warm rhythm of your heart beating against his body, the reassuring rise and fall of his parents' voices, the high, clear sounds of the sonatas and nursery rhymes on which he focused so intently for the first months of his life, will soothe his anxiety and let him know that this world can be a nurturing, supportive one as well.

Over the first few weeks, as your infant's vision develops and the amniotic fluid drains from the ears, allowing him to hear full-spectrum sound, he begins to perceive parallels between the rhythms of this world and the one he knew before. The ticking clock recalls your heartbeat; the give-and-take of verbal communication mirrors the playful interaction of prenatal tapping games, and rocking in your arms reproduces the gentle swaying movement of the womb. Meanwhile, the pattern of light and dark on the wall, the delightful sounds of birds chirping, the delicious warmth of sunlight on his body, all recall his sublime experience of music as it stimulated his senses and informed his prenatal mind. Rhythms, patterns, tones, and inflections are what your infant is born sensing. These patterns of sound, language, and love are the bridge leading him from one world to the next. By holding him close, speaking, singing, and cooing to him, allowing him the physical security of breast-feeding, and exposing him to music he already knows, you can support your baby's exciting transition from a fetal state to infancy.

At birth, your infant's hearing apparatus, consisting of its vestibular (balance) and cochlear (sound-processing) systems, is the most highly developed of his sensory organs, fully prepared to seek out sound patterns that will help the brain begin to integrate the wealth of information that now rushes in. As we have seen, infants are born organically predisposed to perceiving the pitch, loudness, melodic contour, and rhythmic structure of music and of the musical qualities of human speech. You may notice that as you sing or play recorded music for your newborn, he will move his arms, legs, and head in response to changes in rhythm or pitch, a physical sign that he is noticing and storing new aural patterns. By two months of age, he listens to music more intently than to other sounds. He may hold your gaze longer for a lullaby than for a livelier song, showing you that he can intuitively tell the difference between them. At four and a half months, he will know where the familiar phrases end in a Mozart minuet, and he'll notice if the phrases are broken up incorrectly. By five months, he'll sense there is a difference between adjacent keys on a piano, and at six months will show a preference for music composed in the musical scale of the culture into which he was born. By then, his ability to possibly recognize familiar melodies even when the tempo is changed, to recognize a wrong note when he hears one, and to understand music according to the rules of his musical culture will have enabled him to perceive and process music, and many other sounds, in much the same way that you do.

Due to the size of the female pelvis, the human brain cannot grow to its full size before the baby is born. As a result, the brain is only partially developed at birth and will continue to grow *at the same rate as prenatally* for two years afterward. This is perhaps our greatest advantage, and our greatest risk, as a species: that our future growth, our future happiness, depends so much on what we receive from those around us in our first days, months, and years of life. Right now, your baby's heart and mind are wide open, actively seeking your love, support, and guidance. The best thing you can do for him is to hold him in your arms and smile, talk, listen, respond, and *sing*.

A Sound Solution

A BLANKET OF RHYTHM

Pushed out of the womb into a cold, bright, confusing new reality, your baby will surely long sometimes for the familiarity of his prenatal life. You can ease his transition by providing him with the reassuring rhythms to which he's grown accustomed. Studies have shown again and again that during the first few weeks after birth, nothing soothes an infant in his crib like the recorded sounds of the maternal heartbeat, music whose pulse matches a heartbeat's normal pace, or soft classical music that the baby heard frequently before birth.

What works for full-term babies is even more vital for premature infants and newborns in intensive care units. According to a study at the University of California Medical Center, the weight of "preemies" increased faster than usual when doctors played classical music. Another study showed that playing sixty-minute tapes of lullabies and children's songs reduces the hospital stay of premature and low-birth-weight babies an average of five days. Much research suggests that the calming and stabilizing effects of music might help reduce the use of sedative drugs in infants on ventilators and reduce the incidence of respiratory problems.

If the hospital or birthing center where your baby was born does not routinely offer such music to newborns, supply a tape of soft, soothing, heartbeat-related music yourself or, when possible, keep your baby in the room with you, where you have more control over his sound input. Then take your heartbeats home and use them to soften the edges of his first few weeks of life.

PERFECT TIME

"Russian-born Alex spent his first six months in an orphanage, until he was adopted by an American family," music therapist Carol Stock Kranowitz writes about one of her young clients. "Now four, he is healthy, beautiful, and full of energy. VERY full of energy. Indeed, Alex is irrepressible. At preschool, he flits and darts from one activity to another. He does not relate well to other children, who object when he disrupts their play. When his classroom teacher attempts to settle him down with an art project or a lap and a story, he pays attention for a brief moment and then wriggles away.

"Alex's class comes to my music and movement room twice a week. I haven't yet found the musical experiences that will engage him in a positive way. A few days ago, when I strummed the autoharp, he covered his ears and yelled, 'Stop!'

"Today I am teaching an East European lullaby, 'Sleep My Little Bird.' I draw a wooden recorder from my pocket and begin to play the tune. Two measures into the melody, all the children are mesmerized—even Alex. He sits at attention, tilted forward, totally alert. The tune ends. Alex whispers, "That's MY song! Play it again!" While I repeat the lullaby, Alex arises and sways in time to the music. His classmates see his response and scramble to join him. Now the whole group is keeping the beat—rocking from foot to foot, bending forward and back, twisting from side to side—as they play a spontaneous follow-the-leader game. And the leader is Alex!

"What is happening? What has turned Alex on? What has helped him connect to his classmates? Later, I ask his mother if she can explain, but she is as mystified (and delighted!) as we teachers are. Our best guess is that just maybe, his birth mother hummed it to him in utero, or an orphanage caregiver crooned it while rocking him. Something about the tempo, key, and cadence of the lullaby touched a chord deep in Alex's memory, deep in his bones. Something magic."

Indeed, Alex's mind was touched by magic that day in the classroom—the magic of the neurological *critical window*. You already know that before birth and for about eighteen months afterward (after which a necessary culling process takes place), the human brain builds its web of neuronal connections (synapses) at a truly breathtaking rate, and that most of this growth is created through experience, not inherited genes. Cuddle your baby, and an entire area of his brain lights up, forming instant links that will grow stronger every time you cuddle him in the future. Repeatedly fail to cuddle him, and he may well experience developmental delays later on. Learning, in other words, is not an accidental process. It happens as a result of sensory input, and it happens much more easily at certain times than at others.

We all know how difficult it is to learn a foreign language as adults. Yet we look all around us at one- and two-year-olds chattering away—having picked up their own first language with no classes at all and with hardly any apparent effort! This is because very young children are moving through the critical window, or ideal time, for language acquisition. During this period, researchers Peter and Janellen Huttenlocher of the University of Chicago suggest, the sounds of words build up neural circuitry that can then absorb additional words. In other words, stimulation paves the way for understanding, and understanding opens the mind to more stimulation.

Critical windows exist for all kinds of learning. They are the periods during which learning occurs at a much more rapid and effortless pace, and the periods when information laid down is least likely to be forgotten. During the time Alex lived in Russia, his brain was wide open to music—in the midst of a musical critical window—and if a kind caregiver at his orphanage did indeed sing to him in an emotional context, the melody quite likely created synaptic connections in his brain that will last for the rest of his life.

Critical windows can be organic, physiological, cognitive, or emotional, and most of them occur very early in life. If a baby is born with cataracts on his eyes, for example, they must be removed before he is two or he will remain blind even when his eyes are returned to "normal." The critical window for wiring up the eyes comes very early (much of it in the first four months), and if they are not stimulated during that time the wiring (or its activation) simply doesn't happen. Likewise, your newborn needs hugs, melodies, movement, words, and visual stimulation throughout his childhood but will especially need certain stimuli at certain times if his brain is to work up to potential.

Obviously, if this is the case, it is important to know when these critical windows occur, and precisely how to take advantage of them. In this and in each of the chapters that follow, I will focus on the neural areas that are particularly open to rapid and enhanced development. These are the times when certain kinds of stimulation are much more likely to be effective than at other times. Fortunately, most of the ways parents traditionally care for their children nourish their children's brains in ways that are needed at particular times. Hugs, rocking, songs, nursery rhymes, walks in the park, bouncing on the lap and on the knee—these are the simple ways you will help your baby's brain to grow in the ways it needs to, when it needs to. And as we will see, a critical window is a *prime time* to give a child the information he craves, but in most cases it is not the *only* time he can benefit from it.

March of the Neurons

When you first held your baby in your arms and gazed into his wide, observant eyes, you were looking at an already amazingly competent little being. Not only were his basic operating systems—those controlling breathing, circulation, reflexes, and other basic movements—already functioning, but he had already picked up a few higher order tricks in the womb. Among these were the ability to differentiate between people's voices; to hear sound within the pitch and loudness range of the human voice; and to roughly locate objects by sound. He could discriminate among sweet, sour, bitter, and salty tastes. His visual development would take a little longer, but he could already focus on an object an optimal eight inches away (roughly the distance to the mother's eyes while breast-feeding), and would soon be able to follow a moving object with his eyes.

Still he had to a lot to learn, and a need to learn it very quickly. His hundred billion or so neurons had formed no more than about fifty trillion connections. These would have to increase twenty-fold in the *first months* of life, to more than a thousand trillion! No wonder his brain consumes twice as much energy as yours, and will continue to do so as long as he remains a child. The ability to move purposefully is a top priority. By about age two months the motor connections in his brain have grown strong enough so that his initial reflexes (startle reflex, rooting reflex) start to fade and he is able to control his movements a little better. His visual area is also being fine-tuned during this period, allowing him to focus more easily. This is the period of the pensive gaze, when your baby appears so wise and solemn—in short, irresistible.

A very important development process kicks into high gear at about four months and continues to dictate which neural areas begin operating fully all the way through the twelfth year and, in more subtle ways, even later. This is the covering of the brain's nerve pathways with a fatty, insulating substance called myelin. Myelin insulates the nerves, enabling them to speed up the messages that pass along them and thus greatly improve their performance. The fact that the auditory nerve was myelinized prenatally allowed your baby to begin to hear before he was born. As each area of the brain is covered with the substance in its turn, that area becomes fully functional.

By age four to eight months, for example, your baby's visual area has become well insulated. Soon, he will begin to see as well as any adult, though he still won't be able to control his vision for more than a brief period at a time. The corpus callosum (literally, the "great divide")—the thick band of fibers that connects the left half of the brain to the right half—has been myelinized as well by this time, enabling the two sides of the brain to coordinate, and the left and right visual areas to work together. Once this process is complete, your baby will be able to reach for and grab an object.

The background wiring in your infant's auditory area is nearly done by now as well. Your baby has entered a state of high receptivity and listening that will last through about ten months of age. By now, his auditory appetite has become remarkably sophisticated. His taste probably mirrors yours to a large extent. He has learned to love your favorite children's songs, lullabies, and recorded music and has started imitating tiny bits of your speech through his endearing babble. Clearly, this is a very important critical window for encouraging new brain growth through instrumental music, song, rhythm, chants, movement, and plenty of talk. Don't let the opportunity pass to actively nourish your baby with healthy sounds.

The changes you will most likely notice in your little one at around six months are his ability to sit up and his further experiments with sound. He has already created what Dr. Patricia Kuhl of the University of Washington calls an auditory map of your language that enables him to concentrate on the specific sounds you use and ignore, to literally not hear, the ones you don't. Now he will begin seriously putting what he's learned to use, creating new sounds based on the patterns he's picked up from you.

By this time, your baby's web of neural connections has grown astonishingly dense. In fact, there are now one and a half times more branches waiting to carry messages around than your little one could ever use as an adult. In about a year and a half, the excess branches will begin dying off so that the others can work more effectively. Which ones die off and which ones remain depends on the experiences your baby has between now and then. This does not mean you should bombard him with all kinds of stimulation at all hours of the day. (In fact, that's probably the poorest environment you could create.) What you want is to provide a setting in which your baby can naturally explore as wide a variety of objects, events, and sounds as possible, without being forced in any way.

It should not surprise parents to learn that researchers have found that information embedded in an *emotional* context seems to stimulate neural circuitry more powerfully than information presented neutrally. You no doubt remember where you were when you learned you were going to have a baby—but do you remember what was going on around you when you picked up the dry cleaning last month? The point is: if you sing to your baby while rocking him in your arms and making eye contact, rather than chatting mostly with others in his presence or talking to him from across the room, he is more likely to pay attention to your use of rhythm and language. If you respond to his smile by smiling back at him, he is more likely to notice the cause-and-effect phenomenon than if you dully drop a ball and pick it up a thousand times. In general, the more emotional the exchange, the more your baby will pay attention, and the more he pays attention, the faster he is likely to pick up new skills in verbal, physical, and other areas.

Touch is another vitally important route into your baby's developing brain. Holding and cuddling a baby has been shown to bolster the immune system, increase communication, and help premature babies grow larger and healthier at a faster rate. And of course, it's fun, too! When singing your baby to sleep, be sure to hold his little body close to yours. If he holds on to your finger, move your hand gently to the rhythm of your song. Let the vibration of your voice fill him and comforting words nourish his body and brain.

Many of the traditional, natural acts of parenting—playing peekaboo around the edge of the crib, singing and talking as you change his diaper, taking him with you as you run your errands and visit friends, giving him a gentle massage and exercising his limbs while gazing into his eyes, and holding him in your arms at sleep time while humming a soft lullaby—are exactly the kinds of experiences that will stimulate his brain and body. As a parent, you are your child's most valuable tool for building a creative mind, a healthy body, and an open heart.

For Your Child's Health

HE CAN'T BLINK HIS EARS

In the months following birth, it's vital to keep in mind how very sensitive your infant is to sound—much more so than adults or even young children. This is especially true if he was born premature. Severe damage to an infant's inner ear can cause actual brain damage and impede brain development. Though damage to many parts of the brain is correctable at this age, the auditory areas are not among them. Damage to your baby's ears can cause much more harm now than it would in adulthood.

If your baby was born premature and is being kept in an open crib in a neonatal intensive care unit, consider the effects of the twenty-four-hour-per-day sounds of human conversation, medical machinery, rustling papers, etc., on a brain not ready for unfiltered noise. Even in an incubator, sound can be louder than you realize, depending on the ratio of the size of your baby's body to the size of the chamber in which he lies. If his brain and body must defend themselves against an overwhelming amount of noise, that energy cannot go toward helping him grow and thrive. If you are not able to minimize the noise around your baby, at least make sure he is allowed long periods of sleep or inactivity in which to regain his strength. Soothe him with the recorded sounds of your voice, the music he heard in the womb, or a maternal heartbeat—but keep the volume very low.

Once your baby is home, his sensitive ears still need to be protected. Listen to the random sounds that fill his sleeping area. Is there a high level of traffic noise entering through the window? Is the area near a family room, where siblings' shouts or television noise will disturb him? Will a ringing telephone interrupt his sleep and, therefore, yours? Air conditioners and fans can create low-frequency sound that may upset infants. If your baby's room is full of hard surfaces, adding carpets and thick curtains can muffle some of the sounds. A softly played heartbeat tape in the first weeks and gentle classical or folk music after that may counteract some of the effects of the noise.

If the noise level is low enough so that you can comfortably talk over it, but your baby still startles at every sound, he may be temporarily hypersensitive to sounds, due to a stressful birth or simply by nature. If so, you will need to work harder to screen out irritating stimuli: keeping his bedroom door closed while he's sleeping, using a soft, low-pitched voice in his presence, and perhaps even playing soft recordings of such natural sounds as waves, wind, and rain. Be sure to discuss your concerns with your pediatrician, who can offer further guidance in helping your baby manage the aural information that can sometimes overwhelm.

MUSIC APPRECIATION

No one needs to tell a new parent that babies and music go together. From the first moment you hold your child, you naturally feel the urge to rock and softly serenade him. In fact, studies have shown that on a typical day, parents sing their babies through playtime, feedings, diaper changes, a bath, and while driving the car—varying their offerings from play songs to lullabies to a few popular and religious songs, depending on the activity. Why do humans feel this strange need to break out into song in the presence of their offspring? Perhaps their singing represents an instinctive urge to instruct the child in their particular cultural code, to mold the infant's body and brain to conform with the rhythms of his tribe. Or perhaps music's power as a language, its ability to evoke the physical responses and engage the brain, creates such rapt attention in the infant that the parent can't resist making the same connection again and again.

Music's basic components, rhythm and tone, seem created almost deliberately to stimulate the brain and body. Just as your gentle rocking movements, rhythmic taps on your abdomen, and cadence of speaking and singing stimulated your baby's vestibular (balancing) system before birth, so after birth you stimulate his inner ear by rocking him in a chair, bouncing him on your knee, and serenading him with simple nursery rhymes and chants, thus helping him begin to sense where his body is in space. Such stimulation will soon lead to your child's ability to crawl, walk, balance, and move through his days with physical grace.

Tone, meanwhile, can go beyond the stimulation of the pitch discrimination centers and other parts of the cochlear system. After birth, exposing your baby to music without lyrics will encourage his developing musicality, while singing songs lays the groundwork for language ability and, further down the road, reading, speaking, and self-expression.

But, as we will see, music offers a third indispensable advantage to parents and their children—a power that parents have understood and honored since the beginning of time. That is music's potential to convey feelings of love, delight, and security to the baby and its ability to bond family and newborn together in a warm, lifelong embrace.

TUNE IN, TUNE UP

TAKE FIVE

Routines are an important part of parenting, and the sooner you begin creating them for your baby, the sooner he will learn to rely on these rhythms of life. You can take advantage of his critical window for sound by starting a family tradition—a five- to ten-minute musical bath twice a day. At approximately the same time every morning and evening (after a meal is a good time), put a movement of Mozart, Bach, or Vivaldi on your tape or CD player, and hold him close while the two of you listen. Does he enjoy it? If not, try another selection until the two of you find a piece you both like. If the music is of high quality and the same piece is used each time, your infant will begin to develop an "ear" for music at once. Who knows—he may come across it at the piano years from now and surprise you by playing it perfectly from the start!

Music baths should not be the only times for music making. Make a habit of turning otherwise boring chores, such as changing your baby's diaper, bathing him, or putting on his clothes, into brief "sound breaks" that will make the activity more enjoyable and stimulating for both of you. As you lay your baby down to change his diaper, bend over him and smile warmly, holding his gaze. As you smile at each other, improvise a playful, rhythmic diaper-changing chant such as "My baby's so very sweet/How are you, these little feet?/How are you, how are you/How are you today?" Make your voice as expressive as possible, and exaggerate your facial expression in a pleasant, not scary, way. Allow him to participate by gently moving his legs, then his arms, with the rhythm of your chant, or lightly tapping each part of his body as you name it, and end with a gentle tickle on the tummy. Note how raptly your baby studies your performance. His brain is lighting up like a Christmas tree, richly stimulated by this time that might otherwise have bored you both to tears.

A Family in Tune

"When we adopted our four-and-a-half-year-old daughter from Korea," wrote Shirley Besteman, "she didn't speak any English. Fortunately, she was raised by a missionary from the United States (at Che Chon Children's Orphanage in South Korea). She would play choruses such as 'Jesus Loves Me,' 'Jesus Loves the Little Children of the World,' etc., during their daily devotional time. We had received a videotape of our daughter in the orphanage before she came, so we were familiar with the music being played in the orphanage.

"When Kim Soo came, she was very frightened and homesick. My husband would play the 'familiar' choruses on the organ and she would instantly relax and start to smile. It was our way of communicating with her, and it comforted her as well."

Recent brain research has confirmed parents' intuition that warm, responsive care not only comforts an infant but is also critical to healthy development. A strong, secure attachment to a nurturing caregiver not only feels good but also actually immunizes an infant to a great extent against permanent damage from later stress or trauma. Dr. Megan R. Gunnar of the University of Minnesota has studied children's reactions to stress by measuring the levels of the hormone cortisol in their saliva. Cortisol affects a child's metabolism, immune system, and even his brain—by making it vulnerable to processes that destroy neurons and reduce the number of connections in certain areas.

Children with high levels of cortisol tend to suffer from developmental delays in the cognitive, motor, and social areas. Yet some children weather stress better than others—they're simply more resilient. Dr. Gunnar discovered that babies who receive sensitive and nurturing care in the first year of life are less likely to respond to minor stresses by producing cortisol. Even when they do respond in that way, they can more rapidly and efficiently turn off the response. This effect carries forward to later childhood. Elementary school kids with histories of secure attachments are less likely to show behavior problems when under stress. In short, mounting evidence shows that the kind of care infants receive, and the kind of attachments they form with primary caregivers, have a decisive *biological* effect on their ability to express and control their own emotions.

Fortunately, loving our children (and showing them that we love them) is not something that most parents have to work at. In many cases, the process of attachment begins immediately after birth—with parents melting the moment they first hold their newborn in their arms. This state of all-consuming love is called *engrossment,* and it happens to fathers and mothers (and even some grandparents!) alike. As you sing, rock, and lock eyes with your infant, the limbic, or emotional, centers of each of your brains are activated, forging a vital link between you and your child. Attachment is the perfect term for this process, which takes place not just once at birth but continuously over the decades to come, since it serves as the glue to lock families together in a mutually supportive, mutually loving embrace.

Breast-feeding often encourages the attachment process for mothers (as they cradle their infants in their arms and gaze into their eyes), but music can strengthen the connection for both parents, as well as for other family members. In fact, it is one of the best ways for fathers, who may otherwise feel left out of the earliest stages of family life, to connect with their infants. When your baby is fussy and restless and Mom needs a nap, take your baby to the rocking chair and sit with him cradled in your arms. Cuddle him with his side against you and his ear pressed lightly against you chest. Now, start rocking slowly and experiment with a few slow, low-pitched songs until your baby signals by his silent listening that he likes one (a very quiet, soothing version of "We Will Rock You" worked best for one baby I know). The point is to let the baby feel the deep, reassuring vibrations in your chest as you sing. Some babies also like a gentle massage or light, rhythmic pats on the back while rocking. These efforts, plus the slow rhythm of the rocking chair, will comfort him in a way that only Daddy can provide.

All evidence suggests that children are as capable if not more capable of comprehending music's emotional import as adults. A 1994 study showed that musically untrained children were able to identify the emotions evoked by various musical selections as efficiently as adult music experts. This means that listening to music can become a two-way street—a vital means of communicating emotionally for both of you. Certainly, it's a crucial means: stories abound of desperately ill infants whose life support systems were shut off, only to be saved when a doctor, parent, or other relative took the infant in their arms and rocked and sang to him. In any case, no one who has ever witnessed the wonder in a newborn's eyes at the first sounds of a parent's lullaby can doubt that human emotion is a force with which infants are profoundly familiar.

Attachment serves another vital function in the family. It eases the process of attunement within the family—that is, the process of learning one another's physical and emotional rhythms and adjusting oneself to accommodate them. According to psychiatrist Daniel Stern, author of *Diary of a Baby: What Your Child Sees, Feels, and Experiences*, parents and other caregivers who are attuned to a child understand what that child is feeling and can therefore mirror his experience back to him, thus supporting and strengthening his response. When you answer your infant's happy giggle with a hug or a smile, or verbalize his excitement at seeing a dog run across the lawn, you reinforce the circuits for those emotions. Since your baby's brain uses the same pathways to respond to an emotion as to generate one, your reciprocation ("Look! A dog!") reinforces the electrical and chemical signals that produced the original emotion. If the emotion isn't reproduced—if a parent repeatedly fails to notice that her child is smiling and cooing at her—the circuits weaken and may, in extreme cases, eventually die.

Music can be a major force in tuning up your family—not only by deepening attachments through rocking, singing, and enjoying recorded music together, but also by helping the members of your family relax into the rhythms of one another's natures and mirror one another's emotions through dance, playful songs, soothing lullabies, and rhythmic rhymes. One of the unexpected pleasures of creating a family that sings, dances, drums, and enjoys recorded music together is that someday in the not-too-distant future, when Mom or Dad is feeling down in the dumps, a highly attuned child will turn up with a flute or a song to help chase away the blues.

SPOTLIGHT ON THE SPECIALIST

ANN NORRIS OF BRIGHT BEGINNINGS

Ann Norris, M.S.W., is the coordinator of Bright Beginnings, Warm Welcome of Boulder County, Colorado—a free home-visitation program available to all families with children from prebirth to three months. The philosophy at Bright Beginnings is that the addition of a child creates some stress (even if it is welcome stress) for all families, and they deserve support.

The trained volunteers at Bright Beginnings bring a wide variety of gifts to families with newborns, including videos on new developments in brain research, a board book for the baby, and most recently, a recording of Mozart's music especially designed for newborns.

''Mozart's music helps families understand that they can really make a difference in their children's and their own lives,'' Norris says. ''Once parents and their babies start listening to it, they interact on an entirely new level. It uplifts them, and reminds them of the beauty of their lives together. You don't have to be a musician to appreciate that.'' Norris added, ''One of my volunteers told me that playing Mozart seemed so effective that it *almost* made her want to have another baby, just so she could do it right this time!'' Parents as well have expressed their enthusiasm for Bright Beginnings' commitment to Mozart for babies. As one mother put it, ''It's a sign of something going *very right*.''

THE LANGUAGE OF MUSIC, THE MUSIC OF LANGUAGE

Music is magic in its ability to forge connections between the hearts of human beings through rhythm, voice, and tone. Just as miraculous as its ability, once that bond has been established, to lead the infant mind to new intellectual knowledge and expressive ability. The familiarity of the parent's voice—its captivating rhythms, intonations, and variations in pitch—makes an enthusiastic listening process take place. Love, in other words, creates an arena of focused attention, which in turn leads to greater learning and mastery of skills.

Long before an infant can understand the meaning of individual words, he is fascinated by the rhythms and melodies, the musical qualities, in the speech of those around him. The language and music centers of the brain are separate but adjacent, and their development proceeds in a roughly parallel manner. Scientists believe this may be one reason why listening to music seems to stimulate language skills, and why practice with language encourages the active listening that is necessary for creating and performing music. In any case, the innate musical sensitivity your baby was born with focuses his attention on the musical patterns of your speech and the speech patterns in your singing.

Parents usually notice how rapt their babies look when someone holds and talks to them, but watching your baby's response when an adult sings to him may help you see the direct relationship between music and the development of language skills. Dr. Beth Bolton of Esther Boyer College of Music, Temple University in Philadelphia, describe babies' responses in infant music classes as teachers sing directly to each baby. As the teacher approaches, "some babies continue to stare directly at the singer (almost always with an open mouth), entering into what appears to be deep, and perhaps meaningful, eye contact. Some turn toward their parents and burrow their face into a familiar shoulder (showing an emotional response); others respond with changes in facial expression; still others reach out with open arms and hands to touch the singer's face or mouth. All of these reactions are typical of children who are engaged in language absorption and interaction with the significant adults in their lives." In other words, by singing to your baby you are teaching him to listen and to speak.

We have already examined some of the musical qualities your baby has noted in your speech—the *phonemes*, or combinations of sounds, that exist in your mother tongue but not in others. Listening to music before and after birth sharpened his attentiveness, helping him master this skill in the first few months of life. Once he has limited his hearing to your range of sounds, he turns his focus to the rhythms of your language—the metrical patterns that determine where individual words begin and end within your apparently continuous stream of sound. Only after he has learned to separate sounds into words can he begin to learn those words' meaning.

It is amazing, really, how quickly infants manage to pick out these rhythms. Cognitive scientist Peter Juscyk at Johns Hopkins University tells us that by age six to ten months, babies of English-speaking parents develop a clear preference for words accented on the first syllable (as most of our words are). By about seven months, they have an idea of which sounds are likely to signal the beginning of a new word, so are able to separate words from a stream of language and begin to recognize them. Parents of twelve-month-olds are greatly amused by their babies' habit of pointing their fingers and orating like little world leaders—the rhythms and intonations of their babble are a perfect imitation of their parents' speech, though not one of the "words" is real. But what these children have accomplished is a miracle: they've learned all the qualities of speech—the pitch, the rhythms, the tone, the expressiveness—except the semantic value. In short, they've learned the music of language.

Toys for the Ear and Voice

Since your infant has been learning about vocal sounds even before he exited the womb, it is never too early to introduce the joys of language to him through music, song, and conversation. One of the first skills a newborn must master in his new environment, for example, is figuring out where a sound is coming from. You can help him practice this skill by lightly shaking a rattle to get his attention, then moving the rattle to different spots around his head and shaking it again, encouraging him to move his head to better hear the sound. Mimicking his sounds, thus showing him that his sounds attract your interest, encourages him to experiment more with language. Let him study the rhythms of verbal interaction by smiling when he makes a sound and then responding—making eye contact and using plenty of expression, or musicality, in your voice. Other ways to introduce such ideas as turn taking in conversation include playing telephone—talking very expressively on a toy phone, then handing the phone to the baby to "talk" into, then taking it back to talk yourself; by responding to his gestures with sound (for example, repeatedly helping him poke your nose and then making a funny buzzing noise); and by letting him watch you and your spouse sing a song back and forth to each other, alternating lines or verses. (Before long, you will see him move his eyes between you as you sing, like a spectator at a tennis match.) Bouncing him on your knee while chanting, "This is the way the baby rides/Clippity-clop, clippity-clop," will also introduce him to the rhythms of communication, as will playing "Pat-a-Cake," lightly patting his hands as you chant, and "This Little Piggy Went to Market" while you touch each of his toes in turn and end with a tickle. The simple words of children's and folk songs will familiarize him with the cadence and vocal range of his mother tongue. Even the rhymes have a beneficial effect by focusing attention on sounds and rhythm in a way that pleases the senses. Studies have shown that infants can distinguish rhyming verse from unrhymed prose immediately after birth.

Most important, make a point from the very beginning to expose your baby to a wide range of musical styles and genres. Remember, right now he is very efficiently focusing on those sounds that are common in his environment and ignoring those sounds that aren't. What he doesn't hear during these very early months, he may never be able to hear. A lack of exposure to midrange and high frequencies in the first three years of life may also lead to difficulties in listening or learning, so make a point of occasionally playing high-pitched instruments such as the harmonica or a toy xylophone for your baby. You don't need to know how to play tunes; just making a variety of beautiful sounds, not too loudly, will help him begin to discriminate pitch—one of the first steps in learning to decode sentences and words. Naturally, it's best to use well-tuned instruments if you have them available.

Finally, lullabies are one of our culture's best tools for stimulating an infant's language abilities. Their simple pitch contours, elongated vowel sounds, and repeating rhythms are custom-made for introducing verbal communication to your baby. If you sing the lullabies yourself rather than play recordings you've bought, your infant will listen even more intently. Research has shown that adults singing to an infant, rather than alone or into a microphone, naturally sing with more expression and enthusiasm. Your baby will prefer and benefit more from a heartfelt delivery from someone he loves. You can keep your baby's ears open to the sounds of other languages, too, by playing recordings of folk songs or children's songs from other nations a couple of times a week. (This is also a good way to allow adopted babies to maintain their connection to their biological mother tongue.) But then come home to Mozart and the other familiar classics and folk songs of his own culture, so that your baby feels the support and consistency of the music of his world.

The Mother's Voice

The music of a loved one's voice is like no other's on earth. Our voices are as individual as fingerprints—a uniqueness of timbre, tone, and rhythm that caresses the ears of our family members like a long-familiar melody. This visceral response to the human voice is no accident—it is part of a biological process called vocal marking. Like individual scents, postures, or other forms of subliminal communication, vocal marking allows us to tell individuals apart—to identify the members of our tribe, locate family members, and tell friends from strangers in an efficient, nearly foolproof way. Most living creatures use sound communication for such purposes as foraging, reproduction, and social organization. We humans also use our voices to bond emotionally with others, to match babies with their parents, and to enhance our play as well. The minute your baby is born, if not before, you will probably catch yourself instinctively "imprinting" your baby vocally through the high-pitched, musical, silly sounds of baby talk, or what scientists call *parentese.*

Parentese sounds like speech set to music, with its high-frequency sounds, pitch variations, singsong character, rhythmicity, extreme expressiveness, and gliding vowel sounds, all accompanied by dramatic facial expressions. Certainly, it gets babies' attention; since before birth, they have been especially hooked on their mothers' specific voice qualities. The musical aspects of baby talk have been shown to improve infants' emotional and behavioral states, and enhance the attractiveness of nursery games and chants. Mothers can be especially effective in stimulating their babies' language centers by combining an infant's instinctive focus on Mommy's voice with the best qualities of parentese. When talking to your baby, suggests linguist Naomi Baron of American University, speak slowly and clearly, at a higher-than-normal pitch, making it easier for your baby to distinguish individual words. Repeat a word or prolong the vowel once in a while ("We can swing uuuuuup and down!"). Use short sentences at first (though more complex sentences will stimulate your toddler a year from now). As he begins to say his first words, respond to them by expanding on what he says ("Yes! I see the truck!") to show him that he's been understood. Repeating his question in a different way (responding to "Blankie," with "Where is your blankie?") will gradually add to his list of available words.

As you chatter away to your little one, you may be surprised to realize that he is not always a passive participant in this process. He lets you know through eye contact, happy kicking and cooing, and smiles, which kinds of baby talk work for him and which do not. Whether you know it or not, you're responding to his messages—discarding the verbal games and songs he doesn't respond to, and repeating the ones he likes. In this way, not only are you teaching him to communicate but he's also teaching you. As the months pass, and you move from simple words to songs, nursery rhymes, musical games, and stories, your shared repertoire of favorites will grow and grow, creating a pool of lifelong memories.

A Musical Recipe

SAFE AND SOUND

No parent has made it through the first year with their child without wondering how they will get through another bout of wailing. If your baby continues screaming after you've checked him for every possible ailment and other source of discomfort, make a last-ditch effort to soothe him by joining in with his crying. The point here isn't to mimic your baby (infants' feelings can be hurt, too), but to match the pitch of the cry and then turn it into a little song. Once you have your baby's attention, start bouncing him on your lap to the rhythm of the song, or lightly tap the beat on his body. Starting at the pitch of the cry, you might improvise a song such as "Hey, hey, it's okay/You'll be all right, baby boy/Hey, hoo/What's going on with you. . . ." In this way, you can gradually move him away from his sadness and toward your more cheerful rhythms.

"I really had to laugh at the look on his face when I matched his pitch and started wailing, too," one mother told me. "He was so surprised he stopped crying completely for an instant. Then he joined in with me. Then, when I turned the cry into a musical tone and then a melody, he laughed and kicked until I did it again." All proof that sometimes even babies are just looking for a little empathy.

FEELING THE BEAT

It is wonderful to consider how to stimulate babies' minds and beginning language skills, but it's important to remember that music can do a great deal to enhance your child's physical experience, too. In my book *The Mozart Effect*, I described music researcher Howard Gardner's study of traditional music education among the Anang people of Nigeria. Mothers there introduce music and dancing to their infants when they're scarcely a week old, while fathers fashion small drums for their children. When the children reach age two, they join groups where they learn many basic cultural skills, including singing, dancing, and instrument playing. By the age of five, the young Anang can sing hundreds of songs, play several percussion instruments, and perform dozens of intricate dance movements.

During the past century of our own culture's development, our musical sensitivities have been increasingly numbed as we have depended more and more on a technology-based environment. Instead of making music, most of us simply buy and listen to it and, on occasion, dance to it. This passive approach to music eliminates most of our physical experience—our melding with its rhythm, our loss of separateness in its unifying beat. Certainly, all of us suffer from this loss of visceral pleasure, but children and babies suffer the most. Without feeling the rhythm, beat, and vibration of live music, the infant misses his best opportunity to organize his movement, clarify his relationship to time and space, and gain control of the body he is only beginning to get to know.

Aside from its hearing function, the ear's primary function is vestibular—maintaining the body's balance, facilitating movement, and communicating a sense of the body in space. For these reasons (and because it's just plain fun), it's at least as important to engage in physical and rhythmic games with your baby as it is to sing with a good pitch range and to speak expressively. The important thing is to allow your child to feel the pulse beat of the song or chant. You can begin when he is a newborn by playing beautiful music as you gently stroke his back, arms, legs, and head, helping him learn about his body as he experiences the rhythm and structure of the music. Nursery chants are another excellent way to begin the body-mind connection. Place your baby against your shoulder, chanting, "To market, to market, to buy a fat pig/Home again, home again, jiggedy-jig," and lightly tapping his back to the beat—or else sit in a chair, laying him facedown across your thighs, and rhythmically lift and drop your heels as you chant. When he is old enough to hold his head up, try sitting him on your lap facing you, holding his hands and swaying with a simple, rhythmic song such as, "Rain, Rain, Go Away." Dancing in slow, sweeping motions to classical music with your infant in your arms gives him another sense of rhythm and movement.

Bouncing and dancing are not the only ways to introduce rhythm to your baby's body. Tapping on his stomach to a chant such as "One, two, buckle my shoe," and lightly tickling at the punch line ("Nine, ten, a big fat hen!"), and clapping down onto his hands ("Pat-a-cake, pat-a-cake, baker's man . . .") are also investments toward beat proficiency and learning to master the body.

A Musical Vocabulary

You have worked hard to provide your infant with a daily feeding of good music, frequent doses of children's songs and rhythmic games, and plenty of interesting and "musical" conversation. As the months pass, your little one will begin to show you how much he's already learned. Sometime between the first few weeks and the end of the second month of life, he will begin experimenting with vowel sounds, astonishing himself with what he can do with his own voice. At about four months, he'll add a few consonants: first a "ba" or "ga" or two, then next week a new invention. Two-syllable nonsense sounds will follow ("a-ga," "a-ba"), and then, at around six months, singsong strings of consonants (da-da-da-da-da), called babble. By eight months of age, you'll be convinced that his wordlike double consonants (da-da, ma-ma, ba-ba) carry special meaning just for you—and by responding to them as though they do, you will soon make that belief come true.

In fact, the more you engage in this early communication, sharing in your baby's language play and treating it with the respect and amazement it deserves, the faster his language development will proceed. As you talk, chant, echo, and sing, your child will reap quantitative results as well as qualitative ones: infants whose mothers speak to them a lot know, on average, 131 more words at twenty months than those with less interactive mothers. At twenty-four months, the gap widens to 295 words.

If musical ability is also a high priority for you, musical education specialist Edwin Gordon suggests exposing your baby on a regular basis to music without lyrics as well. This will ensure that your child takes note of the musical, as opposed to the lingual, properties of song. Throughout the day, as you celebrate the passing moments with an impromptu song about getting dressed, brushing teeth, riding in the stroller, and so on, don't forget to do a few clapping-and-dancing numbers using only humming, vocalizing, or nonsense syllables. Pick a lovely melody that your child particularly loves. Sing it using only syllables ("Do-do-do-do-dah," etc.) as expressively as you can. Make eye contact with your baby so he can share in your enjoyment. If possible, hold him on your lap and sway with the beat of the song. The point is to let him feel the music without being distracted by words.

RHYTHMS OF DAY AND NIGHT

While you are keeping time to the beat of a Mozart sonata or singing "Rock-a-Bye Baby" for the eighty-ninth time, you can be forgiven for stifling a yawn or two. Your fatigue is the result of another kind of rhythm—the biological rhythm of your child. If it seems unfair that your infant likes to rally all night and sleep most of the day, it at least makes sense: before birth, your rhythmic daytime movements rocked him to sleep in the womb, while the still, quiet nighttime was perfect for exercising his legs.

An important part of learning to be a parent is learning to go with the flow of your child's rhythms whenever possible. Life with a baby would go much more smoothly if we could schedule our entire family's sleeping, eating, and active times around the hours when the baby is sleepy, hungry, and feeling playful. When you can't change your daily routine to match your baby's, you may be able to begin gently helping your infant adjust to yours through a technique called *entrainment.*

Simply put, entrainment is the sychronization of two or more rhythmic cycles. The phenomenon was first discovered by Christiaan Huygens, a Dutch scientist, in 1665. While designing a pendulum clock, Huygens found that when he placed two such clocks near each other and started them at different times, they would eventually end up ticking in unison. Since then, entrainment has been established as a universal physical force in nature that will act on any two or more vibrating bodies as long as their rhythmic cycles are similar.

Entrainment works just as effectively with biological rhythms as with physical ones. Heartbeat and breathing rates, motor movements, and brain waves have all been shown to be entrained by sound. Thus, singing or playing soft, gentle lullabies at bedtime can slow your baby's natural rhythms, lulling him toward relaxation and sleepiness. Repeating the process at the same time each night will put him in the habit of getting sleepy at that time. Eventually, just the first notes of one of those lullabies may start him yawning. Tape players that attach to the side of the crib are now available in stores, and they are an invaluable tool for bringing your baby into synch with the family.

Entrainment can be used to help establish many more habits or routines than just those related to bedtime. Ellen Smith, mother of eight-month-old Connor, plays my *Mozart for Newborns* tape (given to all new Colorado parents through the Bright Beginnings program) during the twenty-five-minute drive from Connor's caregiver's home to her own. The music soothes Connor in the car, she says, and signals to him that they are on their way home. "Connor calms down immediately once the tape begins," she says. "His father is amazed because it works every time!" Julie Brittain, mother of eight-month-old Joplyn Rose and child care provider for Jessica (eight months) and Jack (six months), plays Mozart and a variety of lullaby tapes for all three infants during the day. If one infant is crying, she starts a cassette tape before the other two join in—and instantly, the two get into a relaxed state as the crier quiets down quickly. In other words, all three babies become entrained to the music's calming rhythms. And each time that state is established, it comes closer to becoming a healthy routine.

It is amazing how efficiently and precisely infants respond to musical entrainment. You can use this natural ability to your child's benefit—letting him know through the music you select when it is time to sleep, when to be alert, when to play, and when to move. Rhythm, tone, and melody have already helped your baby begin to attune himself to the patterns of his environment, of your language, and of your love. Now, music's power can gently guide him toward attunement with your daily schedule as well. In later chapters, I will show you increasingly sophisticated ways to use rhythm and melody to move him smoothly and easily through his days and nights. But for now, you can begin to support him through challenging times simply by changing your tune.

A Mozart Musical Menu

- *Variations on Ah! Vous dirai-je, Maman* (K. 265). If you played these variations on "Twinkle Twinkle, Little Star" for your baby before his birth, they are sure to remain one of his favorites. Hold and rock your infant while you listen to the music together (you might sing "The Alphabet Song," "Baa, Baa Black Sheep," or make up your own words to go with the melody), recalling a time when the two of you were one.
- Andante from Symphony No. 25 in G Minor (K. 183). Again, this selection from one of Mozart's symphonies will sound wonderfully familiar to your infant if you played it before his birth. This is not a traditional lullaby, but rather a piece of music that invites you to speak, sing, or chant while holding your baby close. After a few minutes, your newborn will be ready for more restful music.
- Andante Sostenuto from the Violin Sonata in C Major (K. 296). Now the feeling of a lullaby comes to soothe you and your baby. Allow the stress and structure of the day to melt away as the music balances mind, heart, and body. Cradle your little one close to you. Can you feel him respond to the physical changes in your body?
- *Just Heartbeats.* Research has shown that not only are infants able to listen, but that recorded heartbeat and intrauterine sounds are also the most effective in soothing quite fussy babies. The sound of a mother's heart creates a valuable bridge between pre- and postnatal life. A number of recordings of maternal heartbeats are now commercially available.

CHAPTER 4

CRAWL, REACH, AND CLAP

Moving to the Beat (Six to Eighteen Months)

I just followed the sound that I liked.
You just follow your body.
—MILES DAVIS

As your child sits up, balances precariously in a sitting position, and stretches both hands out to seize the world, it might be a good idea to take out Mozart's *Variations* (K. 265) and listen to them again. Not only does the melody evoke the playful teasing of the original folk song, but the very structure of this brief, toylike piece gives us another lesson in how babies and young children naturally learn. After introducing the original tune, Mozart playfully re-creates the melody on a slightly more complex level; reinvents it again with even more delightful rhythmic and melodic embellishments; and then yet again, until the original melody is hardly recognizable in the burst of melodic play with which it concludes. *This is the way children come to understand their world,* Mozart's music seems to tell us—*by picking up a new "toy" (or concept) and exploring its possibilities, turning it over in their hands (or their minds), and playing with it in another way. Let children indulge themselves in the joy of exploration! They'll learn more effectively on their own in this way than from the most well-meaning parental direction and control.*

Over the next year, your determined baby will learn to creep, crawl, clap, and finally clamor her way through the house on her own two feet, all as a result of her overpowering need to *get someplace*—to explore, experiment with, and understand her environment. Most of her attempts at mastery—waving her spoon in the air, climbing every piece of furniture in sight, dragging her pull toy behind her—may look like random play to you, but to her they are very serious exercises in self-education, aimed at fulfilling her physical, cognitive, emotional, and social potential. A young child's play is her work, and you can best support her efforts by encouraging and offering interesting variations on her ever-evolving interests.

At around six months of age, your baby learns to balance well enough in a seated position to reach out and grasp an object. This marks the beginning of an entirely new stage in her life: one of active, self-directed exploration. After getting hold of an object, turning it over, mouthing it, and otherwise investigating its physical qualities, she will start mashing its buttons (if it has any), waving it in the air to see what happens to it, dropping it on the floor to see if it bounces or if you will pick it up, and kicking it with her foot and watching it slide away. If it makes a noise when she pushes its buttons, she will laugh with delight; if she drops it and you give it back to her, her eyes will widen with new awareness. With each variation on her original experience of the object, she understands it from a different perspective, learns a bit more about how it works, and comprehends even more clearly how such things operate in the world. The candylike gratification of this physical, self-directed, active play will inspire her to keep pushing the envelope of learning, experimenting with one hypothesis after another, with the same joy and creativity young Mozart must have felt creating tunes on the keyboard for his dad.

What motivates this relentless drive to explore, experiment, and learn? On a neurological level, the six- to eighteen-month-old's brain couldn't be better primed to process increasingly complex variations on the basic concepts she has already begun to form about her world. The dendritic connections have been growing so rapidly since birth that the brain is now a densely tangled web, with one and a half times more branches available to pass messages along than the baby could ever use as an adult. Those passages that aren't reinforced and strengthened will die off eventually so that those remaining can work more effectively. In the meantime, though, your baby's brain is a seething cauldron of pure potential, stimulated and, on some level, fundamentally affected by each new variation it encounters.

The primary routes your baby will use to feed her brain between six and eighteen months are through *self-propelled movement,* which enables her to grab, bump into, and otherwise interact with objects, and through *listening to language,* which will eventually increase her expressiveness, vocabulary, and sensitivity to emotional nuance and communication. Babies' motor and language development progress at a truly phenomenal rate during this period, and rhythm and sound are absolutely essential to prepare body and mind for this growth. The rhythmic qualities of nursery chants, children's songs, and simple games instill a basic sense of timing in a baby's muscles and mind—an underlying rhythm that will lead to greater coordination, balance, body awareness, strength, physical grace, and, eventually, forethought and the ability to plan ahead. Folk songs and simple melodies will continue to encourage expressive speech, a wider vocabulary, and a sense of well-being and confidence. The fact that all this development takes place in a pleasurable context that also brings child and family together is the miracle of music—and the wonder of the broader meaning of the Mozart Effect.

FOR YOUR CHILD'S HEALTH

YOUR BABY'S EARS

Ear infections are one of the most common maladies to which six- to eighteen-month-olds are prone. Not only are the infections very painful from the beginning, but the damage they can cause may lead to temporary or even permanent hearing and listening disabilities if they are left untreated for several weeks or longer. Fortunately, it is usually easy to tell when your baby has an ear infection. The pain will probably cause her to rub or tug on her ears, and will certainly cause her to cry. She will probably develop a fever and appear rundown and irritable. If you suspect your child has an ear infection—an especially strong possibility if she has recently had a cold or the flu—have her pediatrician check her ears immediately. Don't try to clean her ears yourself, or inspect them by putting swabs or anything else in her ear canal. You will just risk spreading more bacteria and causing greater damage. But do be sure that her ears are routinely checked for infection at each checkup with her doctor. After medication and the doctor's sound advice, remember that your soothing voice and humming may also help lessen the pain for your child.

BABBLE, BOBBLE, AND BOUNCE

In six short months, your little one has grown from a limp-limbed infant who couldn't even raise her head to an alert, competent baby whose motto must be something like, *"Reach, get, explore!"* Once an adorable bundle of uncontrollable reflexes (throwing out her arms when startled, automatically grasping whenever her palm was touched, rooting when stroked on the cheek), she has now learned to control her muscles enough to sit up, reach for, and deliberately take hold of objects, thanks to the continued myelinization of the neurons in her brain. For a while, this ability alone will be enough to occupy her as she learns to manipulate, drop, and otherwise experiment with the objects within range. Eventually, though, her curiosity drives her forward: by about eight months many babies begin to creep, inch by inch, across the floor toward a particularly enticing toy. Usually, though not always, they start to crawl a few weeks after that.

As we will see later in this chapter, matching movement to verbal expression can help a growing child learn to move more gracefully and to link her thoughts and actions. Though your little one is not yet able to sing a song, you can sing one for her as she begins to crawl. Borrow a melody from a favorite children's song and make up your own words, singing them to the rhythm of her movement across the floor. The following lyrics, for example, can be sung to the song "Are You Sleeping?" ("Frère Jacques").

Are you crawling,
Are you crawling,
Little one? Little one?
Underneath the table.
Watch out, it's not stable,
Crawl along. Crawl along.

Though she won't understand your words, of course, she will register the rhythm of the song as it accompanies her exciting new activity. The pleasant sounds will guide her movements, and your happy attention will encourage her to keep practicing her crawl.

As the spinal fibers continue myelinating through the tenth to twelfth months brain and body are able to work more effectively together, and your baby can make increasingly complex movements. She soon begins to pull herself to her feet and begins "cruising" from one piece of furniture to another. It may seem to you that it takes forever for her to move from cruising to actually walking with no support, but have no fear—your baby's physiological development, plus her need to seek out ever more variety in her learning environment, will drive her steadily forward. By the eighteenth month, when she has long been staggering through the house getting into all kinds of mischief, you may well ask yourself why you were so eager for her to walk.

The process of gaining control of the muscles means that your baby will also begin responding physically to music during this year. About the time she begins sitting up with ease, she will begin expressing the way music makes her feel by bouncing or swaying her arms or upper body when she hears it. By eight months, she may try clapping her hands. After her first birthday, she'll vary her responses more—nodding her head to the music, moving her knees back and forward, kneeling and rocking, and babbling—though still not moving in time with the music or singing in tune. By eighteen months, though, she may have mastered the music's rhythm as well, swaying and hopping with her whole body to the beat.

Throughout this period, sound and rhythm can help your child learn to coordinate her body with her brain. By bouncing, swaying, nodding, and hopping, she is able to study the ways her body works, practice moving its various parts in an organized fashion, and even to plan a movement ahead of time and then carry it out. You can use this natural fascination with rhythm and sound to enhance her developmental process and deepen her experience. By offering her variations on what she encounters—say, taking her in your arms and dancing with her to the music she's been swaying to, or interrupting diaper-changing time to "bicycle" her legs, clap her hands, and bend and straighten her knees to the rhythm of your song—you integrate her internal motion with external sound. Later, you will be able to create new voice/body activities, chanting, "Kick, kick, kick your foot/Kick your foot up here" as the two of you take turns kicking a ball; singing, "Up and down and all around/Up and down and all around/Up and down and all around/Here we go today" as she maneuvers her way through an obstacle course of cushions and chairs; and playing at dancing, then freezing, then dancing again as you start, stop, and start your favorite recorded music. In this way, you will continue to stimulate the brain centers responsible for keeping the beat in walking, talking, and sequential thought.

In helping your child explore the workings of her body through music—as with every other aspect of encouraging her development—the best position for you to take is half a step ahead. When she becomes interested in grasping and examining objects, give her objects that make music or sound—baby rattles, musical toys, spoons that can be banged against pans. The variety of sounds will invite her brain to investigate the concept of cause and effect. When she begins to smile and sway to the music she hears, sit her on your lap once in a while and sway with her, or help her tap her knees or clap to the beat. This enjoyable game will give her practice in coordination. When she babbles, imitate her sounds back to her, expand them into a rhythmic chant, and clap and dance or sway to the chant while she watches, or while you hold her. Such chants, along with children's songs and nursery rhymes, work to harmonize your baby's body movements and motor functions by their effect on the vestibular system (the balance center). As your baby gets up on all fours and begins rocking back and forth in preparation for crawling, incorporate her natural movements in your play. If she becomes fascinated with the sound of a shaking rattle, fill a couple of empty plastic spice bottles with rice, give her one, and shake them rhythmically together, making up a chant to accompany your movement. As her language skills develop, start pointing out rhythms that the two of you hear in other places—ticking clocks, machinery, music—and clap and bang and stomp along with her to these.

Most important, demonstrate to your child how much you yourself love music. Put on a favorite recording and dance, alone or with a partner, moving smoothly and freely, mostly with your upper body. Pick your baby up and dance with her in your arms. Or sit on the floor with your baby in your lap and sway to music, patting your thighs or hers—singing along with the lyrics, if there are any, and encouraging her to vocalize, too. It isn't necessary to stick to classical music or children's songs. A wide variety of musical genres, as well as songs in different languages, will stimulate her brain and body, as long as the music isn't too loud. Some popular music feeds beautifully into the one-year-old's obsession with movement, giving her a rollicking beat to move her body to. Make or buy a few musical instruments for you and your child to play with, too—bells, maracas, tambourines, shakers, gourds, and small drums. Don't worry about how musical you are, or whether you're following a particular method or technique. Just get out on the living room floor and have fun making music with your baby.

Finally, when your little one insists on banging on everything in sight, there is no need to take away her instruments. Instead, join in for a while in a little rhythmic improvisation, acknowledging what fun it is, then gradually transfer the spoons or the trash can lids from her hands to yours and continue making music with them as you put them away. Keep reminding yourself that although it can sometimes sound intolerable, random banging is really the beginning of rhythm. If you show her how to turn banging pots into something beautiful, then you may not have to put up with a frustrated drummer in the house in her teenage years.

Studies have shown that babies' motor (and other) development does not take place in a smooth progression but in a series of growth spurts—periods of quick progress followed by times when there is no apparent progress (when your child may be focusing on other skills or just preparing for the next stage of development). On the other hand, the speed and efficiency with which your baby accomplishes such movements as crawling does depend largely on how much she practiced her earlier, preliminary movement skills. In other words, practice makes perfect—which is one very good reason not to confine your baby to a walker or mechanical swing or discourage her from crawling. The more freedom your baby has to move her body in space—to experiment with all the movement variations she and you can invent—the sooner her movement will become relatively effortless, graceful, and efficient. Once mind and body begin to meld in this way, your child's ear is liberated from its need to constantly focus on monitoring balance and the body's movement through space. She can turn more toward other important duties such as hearing, listening, and comprehending words.

A SOUND SOLUTION

OUCHES AND OOPS

If the toddler months are all about movement, they're also all about bumps, scrapes, and other "boo-boos." It's hard for a very young child to understand that temporary pain will ever go away, but a hurt child can usually be soothed by a favorite lullaby—hummed softly while the child is held close to the chest, or played on the child's tape player. One mother I know invented a pleasing song that starts with the phrase "Brush it off." When her toddler falls, she starts singing the song, and her child soon joins in. Together, they "brush off" the ouch. The song refocuses her child so that she deals with the pain more positively.

When it's necessary to apply a stinging antiseptic, tell your child that the medicine will heal the cut, and that humming will make the pain go away sooner. Humming gives the child a sense of control over the experience and makes the unhappy moment pass more quickly.

FIRST MELODIES OF LANGUAGE

"Leslie sounds so cute, babbling away to herself in her crib every morning," the mother of a nine-month-old told me recently. "But sometimes it seems as if she likes babbling too much—like she's permanently stuck in the 'baby-talk' phase. I start to wonder if she's ever going to learn to talk."

In fact, as I told this mother, Leslie was already talking—apparently, a mile a minute. In the world of a very young child, language's sounds and rhythms—the musicality of speech—simply register before words. In English the sounds of baby talk are such phonemes as "ba's," "da's," "ee's," "sss's," and "ll's." (In Japanese they are different—barked "hi's," merged "rr/ll's.") Your baby has been listening to you create these utterances in your own combinations for months. Neurons from her ear have already helped form connections dedicated to these sounds in her brain's auditory cortex. Now she is trying to make the same sounds herself and thus her conversation begins.

By six months, your baby has already internalized quite a lot of knowledge about her native language. She has begun to hear speech not as a blur of sound but as a series of distinct (though still meaningless) words. If English is the language you speak at home, she already prefers normal English words with first-syllable accents. Next, she will begin to pick out the sounds that you in particular use most often.

You may not realize how much you have been helping her in this process. Research has shown that when parents talk to their babies, they adjust their vocal, visual, facial, and physical actions to match their infants' capabilities. In other words, through parentese-type language and behavior, you have naturally been giving your baby just the kind of speech stimulation she can handle, and no more. You have adjusted your speech, touch, and movement activities to suit her preferences. You have experimented with rhythm and pitch, focusing on those your child liked best. You have unconsciously nodded your head to the beat of your baby talk, rocked and stroked your child to the tempo of your singing. As your baby has grown, you have matched the sounds she made, inspiring her to make more sounds, over and over in increasingly complex ways until it has become impossible to tell who is imitating whom. Thus the two of you together have created a unique prelanguage

language, a private duet, that bonds you together and encourages you to keep communicating. In general, the richer your own input, the more you as a parent have indulged in this months-long conversation, the more verbal your child has probably become. Even if she doesn't feel like saying many words out loud just yet, she probably comprehends a great deal. In a study conducted with babies at two, three, five, and seven months of age, 34 to 53 percent of the infants' vocal sounds were part of conversation framed by their mothers' initiating or matching sounds.

It is fascinating to observe how your baby's musical development progresses right alongside her language growth. In fact, if there is plenty of music in your home, your child may well learn to sing before she learns to talk. In any case, at just about the time she begins to create her first phonemes ("ba," "da," "pa"), so, with your encouragement, she begins to sing and babble spontaneously by as early as six or seven months. Soon you will begin to hear her playing with pitch as she talks and sings to herself in her crib. The ability to vary high and low pitches when singing will be among her earliest musical delights. A love of rhythm and joyful bouncing and head bobbing to the beat will be another symptom of the pleasure she takes in the musical stimulation you've offered her.

It is a good idea to allow her to continue experimenting on her own in most of these instances, but make a point now and then of joining in on her activities by offering some Mozartian variations. For instance, if she erupts with a happy "Ba-ba" and smiles when you enter the room, then lean over the crib and babble cheerfully into her right ear, "Ba-ba-ba-ba-ba!" Then, as you pick her up and hug her, continue with, "Be-ba-be-ba-ha-ha-h!" and "Ooohohoh, ooohoh," and a final, "Bababa-ba-bebebe!" As you change her diaper and dress her, make up a song like this one, sung into alternating ears to the tune of "Twinkle Twinkle, Little Star":

Bah-bah left ear,
It's over here.
Bah-bah right ear,
It's over there.
Here comes your left,
Now here's your right.

Listen with both,
And you'll be very bright.
Bah-bah left ear
Bah-bah right
Put 'em all together
And your brain will ignite.

Later, when she's playing with her toys, play Papageno's song from Mozart's *The Magic Flute* (or ask Dad to sing the delightful Pa-pa-pa-pa-pa sounds). Your baby's ears will prick up and her brain will respond to these new sound stimulants.

As with speech, how accurately, richly, and often your child creates musical sounds depends largely on the frequency and quality of the music she hears. Your own singing, with a full spectrum of language sounds, is probably the most important musical stimulant for her now. The combination of your sounds with your emotional "draw" are too much for your child to resist.

By one year, at about the same time that your child begins to take her first steps, she will probably begin experimenting with her first few recognizable spoken and, quite possibly, sung words. By this time, she will have begun linking words to meanings rather than simply enjoying their musical contours. Amid the streams of babble that so accurately mirror your own rhythms and intonations, she will start to name things—kitty, bottle, blankie, nose. Her spoken vocabulary will probably not really take off until after eighteen months, when she's less distracted by learning to walk. Whether your child strikes you as verbal or not at this stage, you can pave the way for more progress through nursery chants and traditional children's songs. Research has shown that, for infants just as much as for adults, music or rhythm helps solidify all kinds of concepts, including words, in people's memories. One study demonstrated that infants just three months old best remembered how to manipulate a crib mobile when listening to the same music that had been played during the learning process. We have all used rhythmic or musical cues, such as the A-B-C song, to remember lists of words or concepts. This same mechanism must be the reason why such words as *cow, dog, cake, farm,* which are so prevalent in children's songs, are among the first words that children learn to speak.

Frequent singing, chanting, and rhythmic play can increase your child's growing vocabulary even as it enhances her motor skills. The best songs are short, simple, repetitive, and sung in a somewhat high but limited range. "Children's songs are an excellent illustration of how a child approaches language," writes Paul Madaule of the Tomatis-inspired Listening Center in Toronto, Canada. "In these songs, the emphasis is on the sound and the construction of words which 'sound' pleasant: they are phonetically descriptive and fun . . . Because they are perceived as games, the child's motivation is stimulated to listen, learn, and vocalize. As a result, children's songs act as a catalyst in this important transition from the infant's nonverbal world to the adult's world of verbal communication. In a way, these songs are like toys for the ear and voice."

By making sure to speak and sing with a wide range of expression, you can make singing sessions with your baby more effective. This will not only improve her own vocal expression and ability to discern meaning in others' words, but will eventually improve her ability to read. It will help her sound out letter combinations on the page and hear them when they're spoken. Dramatizing the actions in songs that have strong images also holds your baby's attention and helps her fix new words in her mind. If you bounce your baby up and down in your lap, or bounce her on the bed as you sing, "Ten little monkeys jumping on the bed," you will cause her brain to fire off neurons in the movement, emotional, and language areas of her brain.

Point to parts of your or your baby's body as they're mentioned in songs. With enthusiasm, chant words in rhythm for her. Use finger play and gestures whenever possible. Sing simple commands ("We all fall down!") and teach your baby to follow them. To emphasize pitch changes, sing songs in pleasingly rhythmic "Ba-ba-ba" melodies as well as those with words. Play with toy xylophones, drums, and bells to help your child internalize the rhythms of language. Read picture songbooks by singing and pointing to the pictures, to increase your child's vocabulary and improve her syntax, semantics, and rhythm (this has been shown to be especially beneficial for bilingual children). Once you have built a solid core of musical games from your own culture, don't forget to continue to expose your child to a wide variety of musical selections in different keys, styles, genres, and cultural contexts, to keep her ears awake and alive.

A MUSICAL RECIPE

"THIS IS THE WAY WE COMB OUR HAIR . . ."

One of the easiest ways to get into the habit of including music in every aspect of your life with your child is to make up songs about everyday activities as you perform them. Invent a crawling song, a getting-dressed song, a going-out-in-the-stroller song, working to keep the rhythm steady and the verses rhyming. If you find it difficult to invent songs from scratch, start by replacing words in simple nursery songs with new words. In this way, you can also expose your growing toddler to a wider vocabulary. By including dramatic pauses, exaggerated facial expressions, and silly made-up words, you encourage expressiveness and demonstrate to her that communicating can be fun.

FASCINATING RHYTHMS

If there is one lesson to be learned from a developing six- to eighteen-month-old, it's that the body can never be left out of the growth process. In this period, when the ability to verbally frame and remember concepts is only just beginning to develop, the brain must be stimulated through the senses. For this reason, writes Phyllis Weikart of the University of Michigan's Department of Kinesiology, "Movement plays a critical role in the formative years for learning and for living." Natural play and movement, she explains, help carve out neural pathways for cognitive development, language acquisition, problem solving, thinking, planning and recall, and creativity. In other words, what we call an integrated or grounded personality—a person who finds it easy to pay attention and concentrate, to create a plan and carry it out, and to come up with new ways to think and to move—can only grow from a child who is on good terms with her own body. Movement grows the brain. It is a vital key to neural development, especially in the areas of memory and higher cognitive function.

These connections between mind and body, between mental concepts and sensory experience, form naturally in the context of simple children's games taught to very young children by older ones (hopscotch, jacks, jump rope, tag, catch, hide-and-seek, musical games), and through such physical activities as walking, bike riding, hopping, and skipping. With these activities, children begin to associate movement with conscious thought and begin to plan their movements beforehand, to recall them afterward and discuss them. Weikart (and other experts) refers to this as "purposeful movement." For example, a child hurrying down the driveway to catch the school bus is focused on reaching the bus. The same child deliberately choosing to skip down the driveway is still going to meet the bus but has consciously included a purposeful movement as well. "When children are engaged in developing their ability to make and carry out a movement plan," Weikart tells us, "to choose a specific movement, to think about the movement while doing it, and to recall and talk about the movement afterward, they are developing the cognitive-motor link and the movement base for learning."

Over recent decades, however, as children have spent an increasing amount of time indoors, often watching television, and less time outside playing, their motor skills have not made the connections with their cognitive development or chronological age. In 1981, for example, one hundred Michigan teenagers and first-grade children were tested on their ability to keep a steady beat—a skill that predicts many other timing-related abilities such as movement and dance, music, language development and reading, sports, and organizing and planning. The results of the study revealed that only 25 percent of the first-grade students could keep a steady beat, as compared to 80 percent of female adolescents and 66 percent of teenaged boys. Ten years later, the success rates had dropped to *15* percent for first graders, *48* percent for teenaged girls, and *30* percent for adolescent males. Only 10 percent of kindergarten children tested at this time were able to keep a steady beat! Since the prime time for developing this basic level of body awareness lasts from birth to age seven, it is safe to say that most of the teenagers who failed the test may feel surprised and frustrated by what their bodies cannot do. They are likely to experience limited success on the basketball courts and soccer fields, feel stymied when trying to master a musical instrument well or express themselves physically, appear generally awkward and out of synch, and end the school day exhausted.

Such a fate can be averted, and, happily, the ways your child can avoid it are universally easy and fun. The crucial concept to keep in mind is the link between physical action and language—combining movement with hearing and thought. When you are playing with your child, name the movements she makes as she does them. Many children sing children's songs that incorporate words and movements. Playfully mimic your baby's movements, describing what you are doing, and then help her imitate your movements as you describe them again. ("Clap, clap, clap your hands . . .") Once your baby has learned to walk, help her experiment with more complicated movements by talking about them and demonstrating them ahead of time. ("Look, I can hop! Can you hop?") Encourage her own creativity and experimentation in movement. The more you and your child move—and, just as important, the more you talk about what you are experiencing through movement and how it makes you both feel—the more your child will learn to "listen" to her own body, to feel attuned with her essential self. In this way, you can help your little one lay the groundwork not only for an integrated, balanced adulthood, but for an intimate, productive relationship with what I will describe in later chapters as the inner voice.

TUNE IN, TUNE UP
TV FOR TODDLERS

Now that your baby is moving more freely on her own, you will probably notice that music on television often causes her to stop in her tracks, move her body rhythmically, laugh, babble, and even dance. It is hard to imagine that this "high-tech" musical experience can be bad for a child when she clearly enjoys it. Yet good parents are right to worry that television can have a detrimental effect on their children. A study conducted by Katharine Smithrim at Queen's University in Kingston, Ontario, has demonstrated that television does indeed reinforce listening and movement skills, assuming that the caregiver takes care to select appropriate musical programs—preferably children's shows that feature simple songs with narrow ranges, repetitive texts and tunes, and enthusiastic television hosts or characters who invite the viewers to participate in the action. On the other hand, researcher Janellen Huttenlocher of the University of Chicago has found that only live language, not television, boosts children's vocabulary and syntax. "Language has to be used in relation to ongoing events, or it's just noise," Huttenlocher concludes. "Information embedded in an emotional context seems to stimulate neural circuitry more powerfully than information alone."

Even neurologically up-to-the-minute television creations such as *Teletubbies* and videos such as *Baby Einstein* and *Baby Mozart* can encourage passivity rather than active learning if used incorrectly. The key to using TV productively, as the American Academy of Pediatrics has recommended, is to keep the time short and actively participate as much as you can.

THE FAMILY THAT SINGS TOGETHER

So here you are, holding your little one in your arms, ready to sing her to sleep. It's just as idyllic as you'd imagined during those months when you and your partner were expecting—your baby's upturned face, open to your every word, your heart filled with love. You open your mouth to croon a gentle lullaby . . . and realize that you don't know the words.

If it's any comfort, you're not the only parent who has stopped, dumbstruck, realizing that you don't remember a single children's song. (The rather hostile "Rock-a-Bye-Baby" doesn't count.) It's always good to compare songs with other parents. The songs at the bottom of each page of this book are simple reminders for your repertoire. But the fast disappearance of our culture's priceless treasure trove of children's songs is no laughing matter. As we become increasingly isolated in our own homes from our own older generation and depend more and more on music recorded by professionals rather than music we make ourselves, our on-hand supply of children's songs has dwindled to a few tried-and-true old chestnuts such as "Farmer in the Dell" and "This Old Man."

Fortunately, families have begun to rebel against this state of affairs over the past decade. A new institution has begun to rise up in place of traditional playground and grandma's-house musical play. This is the neighborhood early childhood music program—Music Together, Kindermusik, Musikgarten, Music for Young Children, Suzuki, etc.—that you have most likely seen advertised in your neighborhood, at your church, or in your local parents' newspaper. Though these programs are admittedly different from the traditions they replace, in a new way they serve the very important purpose of gathering young families together in a joyful musical setting, reminding parents of happy songs and melodies they enjoyed as children and can now reexperience with their kids (or introducing them to folk traditions that they themselves missed), and stimulating children musically in a playful, noncompetitive, nonperformance-oriented manner. "It is remarkable to see the little ones start to get up and move to the rhythm. Even the youngest babies respond to the music," said Fontaine, caregiver for the child of one of my associates. The act of making music together also creates a warmth and positive energy between loved ones and neighbors that gets participants really looking forward to their weekly sessions.

Everyone who participates benefits from group music making, but young children may benefit the most. "You're increasing by 50 percent, maybe more, the number of positive exchanges between child and parent," explains neuroscience educator Dr. Dee Coulter, author of *The Brain's Timetable for Developing Musical Skills*. "The young child hears a different quality of voice in the parent who accompanies them to the music class. The voice shifts from a tenser, faster, reactive base to a more relaxed, contemplative, reflective base. It's going from a harsh and less tonal quality that often kicks in when parents are speedy and trying to keep up with the pace of their lives. . . . The child senses that this is a marvelous state of mind that Mommy just got into, and all you have to do to get her into that again is to sing one of these songs or go, 'ba ba.' And then the mom relaxes and it's great again. So the initiative is coming from the children. The parents aren't going home saying, 'Let's do music.' The children are bringing it up, as if to say, 'Let's get into the state you were in in the music class, okay?' So they're given the opportunity to get their parents back into a state of mind that fosters the bonding and the nurturance that they needed all along." This is wonderful tool for very young children, because this state of coherence is optimal for all kinds of learning at this stage.

Parents also appreciate the developmental advances they observe in their children as they continue to attend music-making sessions. Max and Heather Lloyd, who attend Music for Babies classes with their eight-month-old daughter, Alexa, at the Children's Talent Education Centre in London, Ontario, Canada, report that Alexa now communicates that she recognizes pieces of music, insists on falling asleep to music, and is able to keep a beat. Another parent in the Music for Babies class became an avid enthusiast when his seven-month-old daughter, who had been shaking hands with the other children at the beginning of every music class, greeted him first thing one Sunday morning with a big smile and a handshake!

Early-childhood music programs accept babies as young as newborns or six months of age. Most programs allow no more than eight to ten babies per thirty- to forty-five-minute class, each accompanied by a parent or caregiver. Adults and children sit in a circle on the floor, with each adult responsible for his or her own child, though free movement around the room is usually allowed and even encouraged for the children. Classes usually begin with a relaxing free-play period, during which children can become comfortable with their new environment and parents and caregivers can get to know one another. The teacher then sings a welcoming song ("Clap, clap, clap hello," "Hello, everybody, so glad to see you," "When the sun peeks from behind the cloud, the Sunbeams sing and play"), often naming each child within the song and encouraging everyone to join in. This song is followed by more group singing, during which babies often sit in their caregivers' laps while the adults sway and sing to the music; play with musical instruments; free-form dance; and that of imitation and echoing games. In most classes, the teachers' role is more as guide than instructor: they demonstrate a song or exercise themselves, and the caregivers then show it to the child. All good programs provide musical instruments appropriate to the children's age group; maracas, drums, rhythm sticks, and xylophones are often used.

"I can't believe how much dimension those classes have added to our lives," said Ellen, a new mother who followed my suggestion to join a local group. "Not only did it give Olivia and me a way to meet other families with kids her age, but I could see her respond almost immediately to all the new sounds and experiences. She seems to learn at least as much from the other kids as from us grown-ups. I think she even looks forward to each class's rules and predictable routine. Now she bounces around the room every time I put a Mozart CD on the stereo, and I hear her singing to herself all the time."

Research has shown that continued group musical activities that include singing can increase motor development, verbal abilities, abstract conceptual thinking, and originality in children. Clearly, frequent music making—not just in classes but also every day at home—can make a difference in your child's life, not just intellectually and musically but emotionally as well. Children experience joy, well-being, and love when sharing a class with someone they love. The mother of a baby who attended one instructor's class told her that many times something would happen at home that caused her son to look at her and smile, remembering something similar that had occurred in class. "It is as if there were a little secret that is only ours," the mother said, "and that is very good."

SPOTLIGHT ON THE SPECIALIST

KENNETH K. GUILMARTIN OF MUSIC TOGETHER

As a composer himself and founder/director of the Center for Music and Young Children in Princeton, New Jersey, and creator of Music Together, a national early childhood program, Kenneth K. Guilmartin is fully aware of all the neurological, social, and other benefits to children of participating in a music program at an early age. But what he is most interested in is the great benefit of parent-child involvement as a team. "Children can learn knowledge and how to fine-tune their skills from many sources," he says, "but they acquire the essential disposition to learn only from primary caregivers in the model environment. Therefore, parental inclusion and involvement in the early music learning process seems not only beneficial but necessary, even in the case of parents who present a poor tonal or rhythmic model.'' Music Together emphasizes the value of including parents and other primary caregivers in the music process. Instructors guide parents in learning to dance, sing, and move joyfully to music with their children. Cassette tapes and music books are provided for home use, in the hope that families will get into the habit of making music together while cooking dinner, walking to the store, driving in the car, and so on. The songs on the tapes and in the classrooms are aesthetically pleasing and rely on up-to-date scientific research to provide specific rhythmic, tonal, and language-related experiences for young children. But their main purpose is to inspire families to make music together rather than passively listen to it—to make singing and chanting something you do in life as a part of being an everyday person. This refresher course in the joys of family music making is at least as much fun for the parents as it is for their offsping. One instructor recalls a parent appearing for a class without her child. ''Where's Claire?'' the instructor asked. ''She's visiting her grandparents today,'' the mother replied, ''but I couldn't bear to miss the class myself, so I came alone.''

''True success is achieved when the parents, at least momentarily, forget about themselves and become totally absorbed in music making for their own pleasure. In this state they become the best possible model for the child,'' says Guilmartin. Who knows—even if you've always considered yourself completely tone-deaf, you may find yourself bopping to the rhythms, too, and tuning up your voice, body, and intellect.

RHYTHMS OF THE DAY

It should be clear by now that being a baby or toddler is all about rhythm—how to create it, how to move one's body to it, and how to explore its variations as a way of understanding the world. You can tune in to your child's love of rhythm and predictability by creating a reliable rhythm in her daily routine. Getting her up at the same time each morning, proceeding through breakfast, hair- and tooth-brushing and dressing in the same order, talking about what the plans are for her day and trying to keep playtime, lunch, nap, errands, and bedtime at roughly the same time from one day to another will provide her with a reassuring sense of predictability, and free her up to focus on other areas of life. Each of the transitions between activities, a difficult process for many toddlers, can be smoothed with music. You might invent a tooth-brushing song that you sing every morning, or play a favorite tape as the two of you make lunch together. At bedtime, usually the most difficult transition of all, invent a song, such as the one that follows (which can be sung to the tune of "Teddy Bear, Teddy Bear, Turn Around"), to invite your child into the process of letting go of the day.

Story time, story time
Come along with me.
Story time, story time,
It's your time with me.
Come along and sing a song,
And here we go upstairs.
Story time, story time,
What story would you like to hear?

Mozart can also help move the two of you effortlessly from one activity or state of being to another. To calm your child down after a very lively activity, play Mozart's Rondo—Allegro ma non troppo from Serenade No. 9 in D Major (K. 320). A rondo repeats a melody section of music with variations in between. Your little one will be entranced by the variations and will settle down without even realizing it. Mozart's Flute Quartet in C Major (K. 171) (258b) Andantino is a quiet and lovely piece that will relax everyone in your family. The minuet from Leopold Mozart's *Toy Symphony* will liven things up a bit. The music gets faster and faster so that you can rock, sway, waltz, and play. At bedtime, try the organ arrangement of the *Variations on Ah! Vous dirai-je, Maman* (Twinkle Twinkle, Little Star). The low sounds offer a quiet background to turn out the lights, say good night, and pray your little baby has a long, undisturbed night's sleep.

The period from six to eighteen months is fraught with adventure, excitement, discovery, and even an occasional danger or two. Your child is alive with a sense of variety in every direction she looks. What better gift to give her, along with the alternative perspectives on rhythm, tone, movement, emotion, and expression you have been offering all year, than the security of knowing that some things never change? As the months pass and your child moves from age one to two, give her the gift of predictability through song. She will internalize these pleasant musical messages and use them to move through her days with grace and joy.

DEVELOPMENTAL CHALLENGES

Statistics show that three out of every one hundred babies are born with abnormalities that will seriously affect their health. As babies grow from six months to more than a year, chances are that any developmental problems that previously lay dormant will begin to manifest themselves. You may have noticed that your seven- to twelve-month-old still doesn't recognize words for common items, turn when you call her name, imitate your speech sounds, or use sounds other than crying to get your attention. Your one- to two-year-old may be unable to point to pictures in a book that you name or understand simple questions such as, "Where is your Teddy?" Or perhaps your pediatrician has spotted a problem and has begun to discuss it with you.

Certainly in cases involving hearing problems, Down's syndrome, hyperactivity, cerebral palsy, or even autism, your child's auditory system, with its developmental impact on so many parts of the brain, is likely to figure largely in the therapy plan. Music therapists and other professionals working with developmentally delayed children have increasingly employed music to help correct problems associated with delayed motor development, poor muscle tone, hyperactivity, and sensory processing disorders. In instances where hearing problems are just part of an array of challenges, your child's hearing will (or *should*) receive priority attention.

"Gabe is three years old, and, like many children with Down's syndrome, has a fluctuating conductive hearing loss," writes Katherine Morehouse in the journal *Hearing Health*. "One day, he seems to hear everything and the next, nothing." Adopted at eight months, he arrived at the home of parents Fran and Jay Johnson with a severe hearing loss as a result of untreated ear infections, and no language whatsoever. From the beginning, Fran and Jay took an assertive stance concerning their son's hearing and language development. "When Fran heard, 'Let's wait and see' from a physician regarding venting tubes, she put her foot down and said, 'Let's just do it now,'" Morehouse explains. "When another doctor suggested that hearing was the least of Gabe's issues, Fran changed doctors and pursued amplification for her son. She knew he needed as much hearing as possible to help in development of speech and language. Gabe has been wearing hearing aids since the age of eighteen months, and they have had a dramatic effect on his ability to recognize and imitate environmental and speech sounds. Many children with Down's syndrome have difficulty with oral communication. Even those who are not hearing-impaired often do not have intelligible speech until they reach age four or older. Gabe's parents did not wait around for him to 'not speak.' Instead, they took active steps to insure that his language skills developed in a natural manner through interaction with his family and other children."

Autistic children have frequently experienced great relief through the benefits of music, tone, and rhythm. There is much debate over the cause of autism; it may well have a number of causes. In any case autism, which affects the functioning of the brain, occurs in approximately fifteen out of every ten thousand births. It is a severely incapacitating, lifelong disability that typically appears during the first three years of life. However, with early diagnosis, autism can prove treatable. Recent studies indicate that auditory integration training (AIT), using specially modulated music, benefits many autistic individuals. AIT generally involves listening through headphones an hour a day for ten to twenty days. At Dr. Alfred Tomatis's Listening Centers, autistic children in a number of countries find relief through listening sessions involving high-frequency recordings of Mozart's music and of their mother's voice. Studies suggest that such training may alleviate the symptoms of autism by reducing levels of serotonin, a messenger chemical, in the brain.

Research has also demonstrated that improvisational music therapy improves autistic children's communicative behaviors. The Nordoff-Robbins Music Therapy Clinics in England, Germany, Australia, and New York City provide treatment for autistic and other severely disabled children. The clinics, under the direction of Dr. Clive Robbins, treat patients on a weekly basis in one-on-one sessions or in groups, depending on each child's needs. The Nordoff-Robbins method emphasizes the uniqueness of music as a means of communication for individuals with a wide range of handicapping conditions. Its goal is to "make it possible for developmentally challenged children to reach out beyond the lonely world in which they are entrapped." Dr. Paul Nordoff, an American composer and music professor, and Dr. Clive Robbins, a special educator from England, developed the method together. "There is much these children can't do," Dr. Robbins says. "What we want to know is what they can do."

Rhythmic entrainment intervention (REI) is also used frequently to treat autistic children. REI involves listening to drumbeat rhythms at different pulses or frequencies, so that the subject's brain waves become synchronized with the drumbeats. REI has been shown to reduce anxiety and aggressive behavior in autistic children, and improve listening ability, attention span, and social interaction. The technique can also alleviate the symptoms of attention deficit disorder.

In short, music and rhythm can provide permanent relief for some developmentally challenged children, and an alleviation of symptoms for many, many more. If you discover that your little one is challenged in one of these ways, discuss with her physicians and other specialists the role that sound can take in her life.

A Mozart Musical Menu

- Minuet from the *Toy Symphony* by Leopold Mozart. This is the delightful piece composed by Mozart's father a few months before Wolfgang was born. The "little cuckoo," horn, and glockenspiel call out a delightful tune, inviting you to play "Pat-a-Cake" or "peekaboo" with your baby. You'll hear, too, a rubber ducky and bird join in the musical celebration. This piece is perfect for engaging playtime fun.
- German Dance No. 2 (K. 605). Have your child move her body, stand and sway, and allow the magic of these great German dances to become a part of her play and dance life.
- Rondo—Allegro ma non troppo from Serenade No. 9 in D Major (K. 320). A rondo repeats a melody section of music with

variations in between. Use this selection to dance, gently rock, or quietly play with your baby. You're sure to enjoy this quieter time, too.

- Andantino from the Flute Quartet in C Major (K. 171) (285b). This quiet and lovely piece is relaxing for everyone.
- Adagio (III) from the Quartet No. 20 in D Major (K. 499). Turn the music down softly before you leave the baby's room. The quiet music will mask the other sounds in the house and slowly quiet the mind and body of your baby.
- *Variations on Ah! Vous dirai-je, Maman* (K. 265) performed on the organ. We return to a soothing, slow variation of "Twinkle Twinkle, Little Star" played on the organ. The low sounds offer a quiet background to turn out the lights, say good night, and pray your baby has a long, undisturbed night's sleep.

CHAPTER 5

DANCE AND PLAY

Exploring the Emotions (Eighteen Months to Three Years)

Rings on her fingers and bells on her toes
And so she makes music wherever she goes.
—SEVENTEENTH-CENTURY NURSERY RHYME

When Leslie, a violin student, was three years old, a tape of Fritz Kreisler pieces was one of the few recordings her family owned, so she listened to it over and over. While she was still very young, the tape was lost, and Leslie didn't come across it again until she was fourteen years old and found the tape under her bedroom carpet when the carpet was replaced. Listening to the tape again at age fourteen, she decided she'd like to try playing one of the selections, "Sicillienne." Her teacher, Joanne Bath, gave her the sheet music. Leslie played it beautifully, the first time, in a way so polished it was as though she had rehearsed it for months.

When another of Bath's students, Robert, was two and three years old, he frequently watched a videotape of Shlomo Mintz playing the violin. Eventually, Robert became an outstanding violinist himself. Bath is intrigued and fascinated by how much he resembles Mintz in sound and appearance as he plays.

Stimulated aurally from long before birth, your eighteen-month-to-three-year-old is by now a master listener. You will soon realize this when you hear your least admirable expletives echoed in front of the in-laws, or when he begs to hear his favorite audiotape over and over and over again, focusing on its sounds each time with a solemn, intent expression. Increasingly at one with his body, immersed in a sea of fresh perception, superbly sensitive to the unspoken interpersonal vibrations that fill the air, your child may never be more attuned to the communicative power of music and musical language as he is at this stage in his life. His instincts will serve him well as he enters the world of words. Rhythm, pitch, lyrics, and tone will structure his auditory experience, allowing him to make sense of the babble of language that bombards him from all directions. Increasingly, he'll learn to discriminate among these sounds, not only understanding their meaning but also exploring the nuance, emotional content, and gestures that give them weight. Once he is well along with that task, he can begin to work on articulating expressively himself.

To see just how intimately your particular toddler is engaged with sound, put on a CD of children's songs from other countries while making dinner tonight, and watch him as he listens to this new music. A dreamy expression may settle onto his face. He may hook his arms around the underside of the kitchen table and hang in that improbable posture, raptly listening. If he has heard the song before, he may suddenly break away from that position, hum along for a bar or two, then lie down spread-eagled on the floor, lost to the melody. He may even mimic the sounds of the lyrics in their unfamiliar language. Of course, he doesn't know what the words mean, but he is nevertheless overcome by what they're communicating to him. If he hears this music often as a child, his vivid, emotional, whole-body experience will quite likely come rushing back whenever he hears it as an adult.

The world of emotion is fascinating new territory for your toddler. Of course, since birth he has been sensitive to your emotions and those of his other caregivers, often mirroring your mood or your emotional responses to various stimuli. These experiences created a general emotional pattern for him, a template from which he began making sense of his world. But only now, when he's able to relax his concentration on such physical activities as learning to walk, can he truly turn his full attention to other aspects of life. His eager hands, eyes, ears, and mouth have had the opportunity to touch, see, hear, taste, shake, rattle, and bang on an endless variety of objects, thus introducing an enormous amount of data into his brain. Now his brain takes an enormous cognitive leap; he ceases to be wholly stimulus bound and begins to be able to think about and respond to an object, person, or idea, even if it isn't actually there. This use of a mental image or symbol, rather than an outside stimulus, to drive his actions—say, picturing his ball and deciding to play with it without having just seen it lying on the floor—enables your child to think for the first time in roughly what we consider an adult manner. He is able to remember past experiences and modify his behavior because of them, instead of responding solely to what's happening in the moment. He is able to plan an action in advance and carry it out (even though, of course, his thoughts are still a bit slippery at first). As the weeks and months pass, his emotional life grows increasingly rich as he is able to remember past encounters and interact with others' emotions, rather than just responding reflexively to direct actions.

All of this progress in symbolic thinking is facilitated by this third great leap—an explosion of new language ability. By his first birthday, your child may have been able to say one or two words; beginning at around eighteen months, he will begin learning half a dozen or so new words every day. Language provides the framework within which he can form new ideas, express emotions, and understand concepts. With words, his mind can fly.

Throughout these months, music can work to support, guide, and express your very young child's struggle to comprehend and master the "hidden" worlds he has begun to encounter. Music's regulating rhythms, stimulating tones, and emotional power can reinforce his growth in each of the areas that develop at such a breathtaking rate during this time. Within its reassuring framework, your toddler can practice regulating his own rhythms and begin the long process of mastering a general sense of timing. He can play with words and other sounds, developing his linguistic skills in a fun, natural way. He can explore his own and others' emotions without fear or anxiety. Perhaps most important of all, as he struggles against the inevitable limitations of verbal expression, music reminds him that words are not everything. Emotion, rhythm, tone, and all the other sources of nonverbal information within his body will remain an accessible part of his life.

NEW MELODIES FOR THE MIND

"I can't take it anymore," more than one parent has said to me over the years. "My two-year-old has been asking me the same question all morning. 'What does Little Bear say?' I answer, 'He says, Hello, Mother Bear.' My kid nods and asks, 'What does Little Bear say?' I answer, 'He says, Hello Mother Bear.' It doesn't help if I change Little Bear's dialogue, either. We still have to repeat the same question and answer over and over again."

As any parent who has watched her child open and shut, open and shut, open and shut the closet door knows all too well, repetition defines very young children's behavior. Though there's no doubt that this sort of activity can sometimes make you feel crazy, there are some very good neurological and physiological reasons why your child does it. These months from one and a half years to three are prime reinforcement months, when the overabundant branches in your child's mind will be pruned so that his cleaned-out brain can work more effectively. Unused neuronal connections die out in droves now, while reinforced connections grow stronger. This may sound tragic, but it's a natural, healthy process that helps the brain to function, much like thinning crowded seedlings helps the remaining plants in your garden to grow. When your child repeats a physical action, a verbal exercise, or even an emotional exchange, he is literally building strong pathways of behavior for the rest of his life. This is hard, if largely unconscious, work: by age three your little one's brain will be twice as active as yours.

This great developmental drama begins—both externally, in behavior, and internally, in brain growth—at about eighteen months. When your little one starts saying, "No!" to everything you ask him and bursts into tears of frustration several times a day, you can be sure that his brain is busily myelinating neuronal pathways, thus improving his ability to use mental images—the images that give him ideas about what he wants to do. At around age two, when he begins truly talking up a storm and repeating certain songs, words, and conversations over and over, you will know that the insulation process has reached the band of fibers that connects the speech-production and speech-comprehension areas of his brain. Your child can now begin to coordinate his speech, say the words that he understands, and better handle the rapid pace of conversation. Naturally enough, he now *loves* to talk!

Myelinization is also improving your two-year-old's memory, as well as his fine motor coordination. He becomes able to use expressive language and engage in pretend play. It's interesting to note that 5 percent of toddlers exhibit something called eidetic memory—that is, the ability to remember minute details of a situation. This skill is lost as the child gets older but is believed to indicate a highly developed limbic, or emotional, system.

Equally important, the frontal lobes begin a major growth spurt now that will continue until your child is about six. When fully developed, the frontal lobes act as a sort of executive headquarters for your child's brain. They enable him to perceive patterns, handle complexity, plan ahead, focus, think about the consequences of his actions before doing them, and handle confusion and chaos without panicking. They also allow your child to empathize with others and work cooperatively in groups.

Neural reinforcement, mental imagery, memory, language, expressiveness, pattern recognition, forethought, focus: Music is the natural elixir that can help your young child develop in all of these areas in a natural, enjoyable way. Dr. Dee J. Coulter professes a fascination with the ways in which rhythm, for example, seems to please the mind. "My musician friends . . . refer to pulse, steady beat, syncopated beat, syllabics (or the syncopated beat of words) and rhythmicity" of properly timed music, she writes. Is this the life-giving property underlying the music, the driving force beneath the melody? Whatever its name, it is vital nourishment for the brain. Young children seem to sense this, seeking out and focusing on music as though actively seeking ways to stimulate and organize their minds. Put toddlers in a room with simple musical instruments and they don't just hit and bang randomly on them. Instead, as studies have shown, they play games that deal with the structuring of sounds—creating primitive forms and rhythms, layering two or more sounds, alternating soft and loud, etc. Musically, the children are experiencing and learning timbre, dynamics, texture, pitch, duration, and tempo—the very bases of music. Developmentally, they are gaining knowledge of patterns, connections, and relationships.

As we will see in this chapter, you don't need to have a perfect understanding of these patterns and relationships yourself in order to use music to encourage your child's healthy development. All you need to do is play a wide variety of recorded music for him, actively make music with him in as many fun ways as you can, and open his ears to all the ways in which music can help him grow not only smarter but also more aware and in control of his emotions. Don't worry about trying to choose the most appropriate songs or musical selections for him. He will let you know what he likes, and what he likes is nearly always what he needs. The best you can do is offer him a smorgasbord of high-quality musical experiences, and then take his hands in yours and dance to the beat.

Shake, Rattle, and Roll

What remarkable progress your toddler has made in his motor skills over the past year and a half! He has learned to crawl, to walk, to run, to climb, to throw a ball, and even, unfortunately, to climb a ladder whenever he gets the chance. At eighteen months his movement still needs some practice and refinement. As his physical proportions start to even out, his muscles tighten, and his spine straightens over the months to come, he will delight more and more in his efficient body and its possibilities. Soon, he will learn to catch a ball as well as throw it; build a tower of blocks or even an entire building; and string wooden beads into a necklace for Mommy. He will continue to run, jump, climb, and dance, testing his physical limits and those of the outside world. In the process, he will learn where his "self" begins and the rest of the world ends. He'll develop a better sense of balance, of timing, and of rhythm. He'll learn how to stop as well as start. He'll pick up social skills through simple, movement-based play with other children, and he'll gain self-confidence as he masters each new physical skill.

That is, he will if he's given the chance. Sadly this period, when active movement is so important to the developing child, is the same age when television often enters his life. Instead of playing outside with other kids having a wide range of ages, helping wash the dishes or give the family dog a bath, or engaging in the traditional musical and rhythmic games of early childhood, toddlers are frequently kept safely entertained inside in front of a television set by themselves. Even computer games, which enjoy such enormous popularity among children of this age, may stimulate abstract thinking in some areas but fail to provide the physical experience necessary to teach children balance, coordination, timing, and other vital skills. It is quite clear to anyone involved in child development and education that such physically passive activities, when overused, are a tragedy in the making for children. Movement is an absolute necessity for a toddler, and music stimulates the best kinds of movement.

In 1979, researcher Karen Wolff designed an experiment to determine whether musical activity would have a significant effect on children's perceptual-motor development (skills like crawling, walking, drawing, cutting, the ability to plan and carry out an action). For one year, she studied one class of first-grade children that engaged in group singing and other musical activities, and one that didn't. At the end of the year, she found that the perceptual-motor scores of the first group were significantly higher than those of the second. In the decades following this research, more studies have confirmed her findings: exposure to the rhythms of music and chant can positively affect young children's motor skills.

Music classes are wonderful, but your child doesn't have to attend one to thrive. Likewise, it would be great if kids could still run outside and play all day, but they aren't doomed to lifelong passivity if playing outside isn't a safe option. The good news, as far as toddlers are concerned, is that a little goes a very long way. All you have to do is keep your eyes and ears open for an opportunity to turn a natural experience into a musical, rhythmic one. Make a habit of creating musical moments, moving joyfully to your own drumming and chants, and playing active games with your child whenever you catch him on his feet, and he will be well on his way to the more sophisticated movement play that will nourish him on the playgrounds and in the classrooms of school.

It takes little effort to begin this process, or expand on what you began during his babyhood. When he is sitting on your lap, for example, try lightly tapping his legs in rhythm as you recite or invent a nursery chant. Then bounce him gently on your knee as you continue to speak in rhythm. You are bound to get a giggle or a smile as you turn him to face you, increasing the bounces. He will love the physical movement itself as well as the sense of sharing the beat with you. Touch also stimulates nerve growth. Its power to focus the attention makes it a valuable tool for conveying, anchoring, and integrating information.

If you look, you will find all kinds of opportunities to increase his experience of movement that already interest him. When you play a piece of recorded music, take a second to look at him. Does he bounce to the rhythm, clap his hands on his thighs, nod his head, or sway from side to side with his upper body, beaming at you as he moves? Whatever he does, he'll appreciate your joining in, echoing his movement with your own. As the song progresses you can lead him toward other forms of movement—taking his hands and dancing in a circle, picking him up and twirling with him, slapping your thighs, patting your head, and so on. If you see him swinging his arms to music or even swinging them absentmindedly, start swinging your own arms in rhythm with him, saying "swing, swing, swing." If your child is in the mood and you keep things light and playful, it doesn't hurt to initiate your own movements, too. Get out some scarves and swirl with them. If he's an active child, form a parade through the house and into the yard—taking turns being the leader, of course. Old-fashioned, upbeat tunes such as "Yankee Doodle" or any of Sousa's marches will permeate his body with the structure of rhythm while they instill a warm, happy sense of acting in unison with you.

Show your child that a person can do more than just dance, tap, and march to music. Ideally, rhythm permeates our every gesture, and the sooner your child understands this the better. So put on a CD of dance music while you and your toddler sweep the floor to its rhythm. Wash the dishes to light rock. When your little one is in the bath, give him pots and pans to bang on. Surely, you may have to cover your ears while he practices his drumming skills. If the neighbors complain, just point out to them that moving to music gives kids a chance to develop basic timing, coordination, creativity, and problem-solving skills.

Though most children's songs address a wide variety of developmental areas, some are better than others for encouraging specific kinds of growth. Throughout this period, your child will delight in experimenting with his body's orientation in space and how his body can move and what it can do. Songs that focus on parts of the body prove deeply satisfying to him at this time. Indulge him in singing "One Little, Two Little, Three Little Fingers," "Head and Shoulders, Knees and Toes," and "This Little Piggy Went to Market," and he will beg you for more. As he grows older and more familiar with his physical capabilities, active songs such as "Here We Go Looby-Lou," which involve actually jumping into and out of a space, will also help work out his energy.

All toddlers could use help learning to stop a movement (which is a much more difficult skill than starting a movement). Don't give up the freeze game you taught him when he was younger. Play music, then suddenly stop it as a signal for him to freeze in place. If he is practicing motor sequencing himself through one of those admittedly annoying, repetitive open-and-closing-the-door movements, turn your irritation into music by chanting along with the activity, "*Push* the door *pull* the door *push* the door . . ." until he has integrated that movement and moved on to another.

Research has shown, again and again, that active involvement with music is more effective in nearly every way than passive listening alone. When that active involvement springs from natural movements that children create themselves, their music-related movements increase and their musical perceptions are enhanced. Moving to rhythm with your child will help him harmonize his movements and motor functions, increase his awareness of how his body works, improve his coordination, and help create a positive body image, all while providing much more fun for everyone than another hour of TV.

MUSIC TO THE EARS

Think again about your child's response to the music tapes you played while you cooked dinner. You probably felt the same wash of intense feeling when you were a child and perhaps reexperience it when you hear that same music as an adult. Music, like scent, seems to directly touch the emotional and memory-related centers of our brains. This feeling connection, so deeply intertwined with our physical and sensory selves, provides the hook for all kinds of music-inspired verbal and nonverbal growth in children, stimulating the listening skills that lead to better expression.

Clearly, language is not just a matter of vocabulary. Your two-year-old can listen to an entire anecdote related by one of your adult friends, be familiar with most of the words, and still ask you what it means. His insistence on repeating conversations, described earlier in this chapter, is a part of his effort to plumb the depths of your language to find out what's underneath. Music without lyrics, particularly the expressive music of Mozart, can satisfy your child's craving for meaning even before he learns to say very many words. Music and songs from other cultures can keep his ears alive to the many, many variations in sound and meaning that aren't always expressed in our own music. Make a habit of sitting or lying down with him once or twice each day, at naptime, if possible, or during the times he used to have his musical bath and listen to music together. Don't talk during the selections. Afterward, if he likes, you might discuss how they made each of you feel. But focus on allowing this to be a sharing time for the two of you, a time to simply soak up the nonverbal aspects of communication.

As the months pass, your little one will become fascinated by the concepts of "same" and "different." These are not only cognitive concepts, they are aspects of language. The ability to note differences in sound, facial expression, and gesture can make all the difference in a person's ability to communicate effectively. Around the time of your child's second birthday or a little later, you can begin exploring this idea through musical games that emphasize the sameness or difference of music pitch, rhythm, intensity, duration, and timbre, all of which are qualities of the human voice as well. For example, you might strike a note on the piano or xylophone and then sing either the same note or a different one. Your child can tell you if the sounds are the same. If he's right, play a fanfare for him on the piano. If you can't do that, make a playful "ding ding ding" game show–type sound. If he's wrong, laugh and do it again.

As always, nursery rhymes and children's songs, such as "Here We Go Round the Mulberry Bush," will stimulate your child to actively listen, learn, and eventually begin vocalizing himself. Now, though, they will also offer him a wealth of new words presented in a format that will help him remember them. As your little one moves from age one to two to three, he will delight in memorizing the letters in "The Alphabet Song," and the numbers in "One Little, Two Little, Three Little Fingers" and "There Were Ten in the Bed and the Little One Said . . ."

As you teach him to count the steps rhythmically as he climbs the steps to Grandmom's front door, or to recite his address in a song you've created, be sure to comment admiringly on what a big boy he is, learning such big-kid skills. For eons, children's songs and chants have taught little ones the basics of vocabulary, reading, and math while showing them that learning can be fun and fulfilling. Music is a learning tool that never stops working.

In general, the more dramatically you sing, the more actively your child will listen and the more he will be likely to retain the words. When singing "Five Little Monkeys," for example, vary your voice from a sturdy call to a whisper, act out the movements, and end with a great big smile and a hug. You might use your child's puppets to help sing the song. Accompanying yourself and your child with simple musical instruments, such as xylophones, autoharps, bells, or rhythm sticks, will stimulate his listening skills. As he reaches the end of his second year, he will love the feeling of unity achieved when the two of you sing together. To help him enjoy this process, start by singing the difficult parts yourselves and encouraging him to join in on the chorus. When he's older, you can switch roles. Finally, to help him practice with his own expression, play "echo games"—singing one phrase of a favorite song, letting him sing the phrase back to you, then singing two phrases and having him echo them.

As any toddler could tell you, if he only had the words, human communication begins not in speech but in the body. For true communication to take place, that physical sensation of language, or *felt sense*, must originate within the speaker and echo within the listener. Sadly, though, many adults have forgotten to incorporate their bodies in their everyday speech. They speak, even to their children, in a monotonal series of words that convey dictionary-type meanings but don't make much other sense. In other words, they have speech but no voice. So the young child is not given a voice to listen to. He has nothing to connect to with his own physicality, and so he misses the opportunity to fully listen.

There are a number of ways to keep your child's ears awake through the rhythmic qualities of human speech. Rhyming is an example of speech that touches us physically and thus calls our attention to the sounds inside words. In addition to sharing rhymed children's songs and chants with your child, read him plenty of books with text that rhymes (Maurice Sendak's *Chicken Soup with Rice* and all of Dr. Seuss' books are perennial favorites). Tell him, "Rhymes are fine! Words that rhyme, rhyme in time. The time is fine to make a rhyme, Rhyming rhymes is ice cream time!" Once he begins to recognize rhymes, he will crow with delight when you rhyme your speech for him, and delight in picking out rhymes he hears in the talk all around him. Many a car ride has proceeded more smoothly by keeping the little ones busy thinking up more rhymes.

Physically experiencing the rhythms of words continues to help your child develop a felt sense of language. Knowing where sounds of words begin and end is also important in eventually learning to read. In addition to encouraging your toddler to move rhythmically to nursery songs and chants, you can clap out the beat of individual words that he knows and uses. For example, while you're sitting around doing nothing someday, say, "I can clap out my name—Mama." Then say "Ma-ma" and clap on each of the two syllables. Next, ask your child to clap the rhythm while you say it. If he's very young, you might take his hands and help him at first. Next, clap out your child's name, either together or taking turns. Finally, try other favorite words, such as *ba-by, ice cream, base-ball, ba-na-na, ap-ple, bat, bat-boy,* and *bas-ket-ball.*

Once your toddler has had plenty of experience with feeling the rhythm of words in his body, it's time to up the ante a little. While he's still using simple three- and four-word sentences, move away from your simple parentese and toward more complex conversations with more interesting rhythms and increased turn taking between the two of you. Craig Ramey, professor at the Civitan International Research Center of the University of Alabama at Birmingham, calls this the "exquisite dance" that good conversationalists engage in: knowing when to speak, when to pause, when to listen. We will explore this concept more in later chapters especially as it begins to affect social and academic skills. Now it's important to begin leading your child along a bit in language complexity. Don't just play with music and language, in other words; talk together afterward about what you did!

A MUSICAL RECIPE

IMPROVING PRONUNCIATION

Most two- and three-year-olds are hard to understand once in a while. Their odd accents and speech impediments usually clear up around age four. The improvement comes once the child has listened to the correct pronunciation enough times to be able to reproduce it. However, if your three-year-old's speech is so hard to understand that you feel it's seriously getting in the way of his ability to communicate with others, you might play a few listening games to speed this process along. If, for example, your child tends to say "dute" instead of "cute" (as in, "Aren't you dute"), speak clearly toward one ear and say, "cute, cute, cute," raising or lowering your pitch a little each time you repeat the word. Do the same toward his other ear, again using different pitches. You may soon notice a marked improvement in his pronunciation of that word.

We feel the rhythm of language in our bodies; we feel its lyrical qualities, too. Our complete felt sense of words not only stimulates our musical sensibility but, as we will see in later chapters, eventually leads to the development of complex thought, subtlety in distinguishing among ideas, and the ability to conceptualize on a higher, holistic level. In other words, fully expressed speech fine-tunes the mind.

Dr. Coulter suggests doing the following experiment in using sound to draw in and educate your child. First, stand up and say the following two lines: "Over the rainbow and under the bridge. Through the tunnel and around the ridge." Now add dramatic gestures to the words and speak in an evocative way: "Ooooh-ver the rainbow, Unnn-der the bridge, Throooough the tunnel, And arooouuund the ridge." Do you feel the difference in your body as you say the words first one way, and then the other? Your child feels the same difference when he hears them. Which listening experience is likely to be a richer one for him?

Like the tip of an iceberg, the dictionary definitions of words carry only about a tenth of their meaning. The rest resides in the melody of speech—the rhythms, lyricism, and timbre. In his effort to learn how to communicate most effectively, your child is focused on studying your voice as much as he possibly can. So next time you read or tell him a story, try using a more expressive style. You don't have to be so extreme in your gestures and expressiveness that you look and sound ridiculous. The point is to use language's rich melodies to enchant, stimulate, and engage him in the brand-new universe of words.

Another technique to help him fully understand the complexity of language is to record an audiotape of yourself reading to him or having a conversation with him. Add more sessions to the tape until it is at least half an hour long. Let him listen to it alone at naptime or bedtime, when he's alone in his crib or bed and can absorb the music of your interaction. He will soak up every nuance that passes between the two of you and, most likely, the reassuring messages he picks up will lull him into a blissful sleep.

Rhythm and tone reside not only in musical compositions and human speech, but in the general environment as well, if we only listen. Unfortunately, ours is not a culture that encourages active listening. The sheer level of noise all around us causes everyone, including children, to turn down their level of hearing simply as a defense mechanism. To encourage your child to listen to his world, even when no one's talking, take some time to tune in to the sounds that reach your own ears. Do you hear cars passing nearby, birds singing, fluorescent lights humming? Can you hear your own breathing? Once you have sensitized your own ears and, if necessary, turned down the noise in your environment, you can begin to help your child appreciate the music in everyday life. Better listening will lead to better language skills, more efficient mental processing, and the chance to be fully in tune with ourselves and with others.

It's easy to start your child listening to his world, just point out the sounds to him. When you go for a walk, see how many sounds the two of you can identify. Sitting in the park or in a restaurant, close your eyes and name all the sounds you hear. At home, have your child close his eyes and create a variety of funny sounds for him to identify. Remark to him how much you love the *sssh* sound of a running shower, or imitate the rise and fall of a distant siren. Talk about sounds with your older toddler, and compare them when you hear them, as in, "The tuba on this tape has a low sound. What else makes a low sound? Can you make that sound?" While driving in the car, play "turn off" and "turn on" your ears. Whatever the activity you choose to stimulate your child's ears and mind, joy, not instruction, must be at the heart of listening. When joy comes first, focused concentration and even rigor will follow.

Spurred on by the sounds he has explored, your child will eventually start to ask himself, "How many things can my voice do?" Each day will bring him new discoveries relating to the expressive capabilities of his body. He will learn to mimic a dog whining, a bird tweeting, a cowboy singing, all to enhance a story he is telling you. He will learn that he can sing in a range of pitches, and with practice can extend that range. He will experience the joy of teaching familiar songs to others and learning new ones. Someday, he may even actively explore a musical instrument or two, and use it to broadcast to the world just who he thinks he is.

A SOUND SOLUTION

NOISY TOYS

Exposure to noise is the leading cause of hearing loss in over twenty-eight million people in the United States today, and there is increasing evidence that hearing loss is occurring at younger and younger ages. Six percent of fourth graders, 9 percent of tenth graders, and 61 percent of college freshmen have a measurable hearing loss. Younger children may be even more at risk, since a small child is likely to hold a noisy toy close to his ear. Children raised in noisy homes have been shown to develop cognitive skills at a slower rate than those raised in quiet homes. Environmental noise can also lead to higher-than-average blood pressure, poorer reading ability, a decrease in social interaction, and lower self-esteem.

Since a young child is not likely to tell you about the slight muffling of sounds in his ears, a ringing noise, or difficulty in understanding speech that are the symptoms of early hearing loss, the best way to protect him is to avoid overexposure to noise in the first place. Do your best to keep him away from firearm toys, firecrackers, cap guns, loud recorded music, loud horns, and very loud squeaky toys. The best test for a toy's loudness level is your own ear: if a toy held close to your ear sounds too loud, it probably is. To avoid harmful effects from environmental noise, teach your child to cover his ears when he hears an ambulance siren, jackhammer, or other loud noise. Make sure he learns never to shout in anyone's ears or to let them shout in his ears, and avoid shouting yourself. If you suspect that his hearing has been damaged, talk with his pediatrician. The earlier such problems are diagnosed, the better the chances that they can be successfully addressed.

A BRIDGE OF SOUND

"Reyna Rose was two years, eleven months old when we adopted her from Russia," writes professional musician Debbie Fier. "She was fluent in Russian and had never heard English before. She had been in a children's home most of her life. She had been taken care of better than many in that setting, given that there were forty to fifty children and only four or five adults. But she had not had parental love—including being carried or held, playing with baby things or being taken places in a stroller. She arrived as a small, anemic-looking toddler. . . .

"Since she was almost three years old, Reyna's personality had already been forming, and we needed to begin to get to know one another. Because of our language barriers, it was sometimes hard for me to know what she was trying to say. One of the ways that I moved toward her and let her know that I was trying to understand her was through my voice. When Reyna would get upset, angry, or frustrated by my not understanding her, she would vocalize nonverbal sounds such as 'AAHH.' What I began to do in these situations was to respond to her sound with my own sound, for example, 'OOHH.' Then she would respond to me, and on we went, having the conversation through sounds that we were unable to have with words. What I noticed was that through our 'conversation,' Reyna felt reached to by me and was able to release her feelings of anger and frustration. Other times, particularly if Reyna was sad or upset, I would actually match the vocalization/nonverbal sounds she made. If Reyna was crying, I would vocalize with her at the same pitch that she was crying on, and often her feelings of sadness transformed into her feeling comforted and attended to. This was a way for Reyna to feel reached to, held, and supported by me. She often landed in my lap. Using nonverbal communication brought her back to her preverbal ages and allowed us to connect as infant and mother, although we hadn't actually spent those early years together. She began to feel more of a maternal connection and to embark on a new way of being close to other people that she had not experienced before."

To a very young child, emotions must sometimes feel like a tidal wave, sweeping everything away in its path. Without the experience or sometimes the cognitive development to understand what is happening to them or to control it, without sufficient language skills to express how that makes them feel, they have no choice but to get caught up in the passion of the moment, screaming over an apparently minor incident, sobbing over a sad song, or giggling uncontrollably as their parents make silly faces at them. It makes sense, then, that one of your child's primary missions at this age is to try to understand and manage his own and others' emotions. To some extent, his brain helps him toward this goal: between ten and eighteen months of age, the emotion regions of his brain become linked to the rational prefrontal cortex. The circuit seems to grow into a control switch capable of calming agitation by infusing reason into emotion. By making a point of tuning in to your child's stress and responding to it in a rational, soothing manner rather than threatening or becoming angry yourself, you can strengthen the neural connections between the rational and emotional parts of his brain, and help him learn to calm himself down.

Of course, this is not as easy to do as it sounds. Between eighteen and thirty-six months, he will proceed through all the emotional steps of being a toddler, from the constant "no's" through the tantrums of the terrible two's to the inexplicable sobs and sudden aggression of the older months. Thank goodness his emotional highs will be just as intense. Watch the joy with which he greets his best friend on a play date and how passionately he hugs her and smiles into her eyes. Note the priceless tenderness with which he decorates Mommy with lovely scarves, ribbons, or whatever comes to hand, even if she's apparently ignoring him and talking on the phone. During this period, when words are not always available, music can offer positive ways for him to explore, express, and learn to release his emotions.

"I want to share a story from my occupational therapy practice that happened many years ago but always stays with me," writes Susan Tobin, a former student of mine. "I had a two-year-old child who was nonwalking and nonverbal. When I worked with her I would sing and chant to keep her happy. Otherwise she would cry and pull away when I tried to help her move into weight-bearing, antigravity positions. One day I was chanting 'I can do it, I can do it, I can do it, I can do it' in a singsongy way. She looked up and repeated 'I can do it' and began babbling. She began cooing and playing with sounds after that."

Self-esteem, the feeling that you can do practically anything you set your mind to, is one of the most basic emotional "attitudes" that are formed through myriad unremarkable incidents throughout early childhood. Certainly, high self-esteem can make it easier for a child to handle the emotional challenges that await him. As parenting expert Dr. T. Berry Brazelton points out, your child has been testing his own importance in his world practically since the day he was born. Back then, almost every time your baby smiled, someone smiled back. Nearly every time he vocalized, someone responded with sound. Now, as he accomplishes a task and his face lights up as if to say, 'I did it!' you give him a hug and a kiss. Each of these experiences adds to a sense of trust in his world and in the future. Once that sense of trust is established, he feels safe to explore the sometimes scary emotional world.

As always, the best way to let your child know you're there for him is to listen and respond in a more sophisticated way. When your little one is in a particularly good mood, or the two of you have set out on a walk on a beautiful day, encourage him to "Do a happy hop!" When he's sad, hold him in your lap and let him feel the comfort of your rhythmic swaying and vibrating voice as you talk with him about what's wrong. Even his fear can be at least somewhat alleviated through song. Dentist Robert Wortzel of Summit, New Jersey, has created a videotape demonstrating how children's natural aversion to dentistry can be both expressed and reduced through humorous songs. In short, your child's developing language skills make it easier to help him control his emotions, just as his experience with rhythm and movement helps him express them.

Music can be used to stimulate emotional involvement as well as to soothe overwrought emotions. When you're driving in the car and it's clear your child is bored, for example, invite him to join you in some silly, easy-to-sing nonsense songs. He will soon be laughing at your lyrics, repeating them, and perhaps even making up some of his own, his fretful mood almost instantly transformed. After naptime, an afternoon period when he may just be knocking about the house, put on a recording of *Oklahoma, The King and I, South Pacific,* or any other big, booming, blustery musical. Then linger nearby, available but not intrusive, while he dances, sings, or emotes his heart out. It doesn't matter if he doesn't understand the words, he will understand at least some of the emotional content of the music, and that will be enough to set his imagination on fire.

In fact, leaving your child alone with music, whether it's recorded selections to which he's listening or musical instruments with which he's experimenting, is one of the best ways for him to learn to express his joy. A number of inexpensive instruments make perfect experimental tools for a toddler. Try small hand drums between seven and twelve inches in diameter (the smaller the child, the smaller the drum should be); paddle drums (whose handles are useful for children with physical disabilities), small sets of bells, with which he can explore differences in pitch; and Autoharps and Chromaharps, which emit chords when your child presses the buttons or strums the strings. Appropriate musical instruments are available through early-childhood music programs and music supply stores. (See the Sound Resources section of this book.) But your child doesn't need professionally made instruments to make music. You can help him make a drum out of an oatmeal box and a cymbal out of two cooking pot lids. The point is to get him playing.

Once your child understands that you accept his emotions and are not frightened or dismayed by them, he will become curious about your own emotional responses as well. Children learn mostly through imitation at this age, so this is your chance to model the emotional experiences you hope he will learn to have. To do this, choose some of your all-time favorite songs, those that most tug at your heartstrings, and sing them to him as expressively as you can. It will be hard not to laugh at his little face mirroring your overdramatic expressions, but try not to. He is learning about emotion as he studies you and deserves to be treated with respect. It's also fun to listen together to expressive, high-quality recorded music that you particularly like. Tell him, or act out, what you like about the music even if he's too young to completely understand, tell him what your favorite parts are, encourage him to hum or move along with the music as you do, and ask him how it makes *him* feel. As he approaches age three, consider treating him to some live, child-oriented musical performances as well. Watching a guitar player's face as he plays his instrument or observing an orchestra conductor's body emoting as she leads a symphony is a profound musical, emotional, and aesthetic experience for your child. The love of making music is an emotion that you can respect and that all children can and should learn to understand.

Another way to share your own emotions with him is to make music together yourselves. You might start by dancing to the beat as he pounds on his toy drum, or drumming yourself while he dances and moves around you. If he's old enough, calling out directions such as "Freeze!" "Faster," and "Go backward!" adds to the game's complexity. If you have two drums or two kitchen pots turned upside down, play call and response with each other: you drum a rhythm, and have your child respond. Then you respond to, or echo, his drumming. Drumming and moving to rhythm are also wonderful ways for you and your child to release excess energy. Your child will appreciate your help in learning to manage himself in that way.

As he gains confidence in his understanding of his own and your emotions, your older toddler will begin to turn increasingly outward, exploring what it is to be like a dog, a horse, a car, or the lady next door. These elaborate explorations of others' emotional lives are not only evidence that his imagination is expanding, but are also a sign that your child is developing emotionally in a healthy way. Encourage these forays into the subjective world of others by encouraging your little one to "sing like a siren," "talk like the postman," or "dance like a cat." Experimenting with what it's like to be someone or something else leads to the first feelings of empathy, which in turn lead to a moral sense. Listening to one another, expressing one's feelings and trying to imagine another's, and learning to take turns in musical games and elsewhere are all precursors to behaving according to the golden rule. Only by supporting your child through all the stages of emotional development will you teach him to one day be compassionate.

SPOTLIGHT ON THE SPECIALIST

LORNA LUTZ HEYGE OF MUSIKGARTEN

In 1994, Lorna Lutz Heyge, director of the Foundation for Music-Based Learning, resolved to create a new kind of early music program that would emphasize the young child's developing emotional connections to the world around him. Aware of the plethora of research demonstrating that structured group music instruction increases vocabulary and language ability, motor development, particularly coordination, abstract conceptual thinking, play improvisation, originality, and social interaction in young children, she remained primarily focused on music's ability to bring families closer together, to help the child explore his inner self, and to offer children the freedom to be themselves. "I look for the innate music in all movement," Heyge says. "Every *body* is musical."

At present, the program, which is available across the United States, offers a Musical Bonding program for newborns to children aged three, which introduces adults and children to the songs of their culture; a Cycle of Seasons program for children aged two and a half to four and a half, which draws its central themes from nature and involves both children and parents; and a Music Makers program for those aged four to seven that emphasizes the training of the ear and voice and body movement, which gives the child a solid foundation in making music.

"By using songs that reflect the seasons, the children's inner and outer worlds begin to blend," says Heyge.

''Moving with the trees and grass, or watching a leaf fall and then rest for a moment on the ground, creates a wonderful awareness of movement, flow, and rhythm.''

Much as he has studied the world of emotions by the time he reaches age three, your child can still be thrown off balance, as can we all, by an onrush of unexpected emotion. You will provide a valuable service for him if you can find ways to use music to help him through these difficult periods. If nightmares have become a problem, or if your child is afraid of going to bed alone, it might help to create songs together during the daytime about fears your child has at night. In this way, he can confront his fear when it's least threatening and explore ways to conquer it. On those occasions when your child calls to you, frightened, at night, hold him in your arms while showing him that simple humming can distract him from his anxiety, filling his body with a reassuring vibration that leaves no room for "monsters." Familiar music can also act as a security blanket when a child is frightened or anxious because his sleeping routine has been disrupted by travel or some other way. Be sure to pack his favorite Mozart recordings when you go on overnight trips. At home, transfer his crib-side tape player to his toddler bed so that he can learn to comfort himself in the middle of the night.

As the months pass and your child's unique character unfolds, you may begin to wonder whether certain emotional characteristics are less the result of moods or experience and more just an expression of your child's basic nature. In fact, your child has inherited a specific physiological temperament through his genes, a tendency toward shyness or aggression, for example, or an extroverted or introverted nature. Though research has shown that it isn't possible to completely cancel out an inherited temperament, it is quite possible to coax behavior away from its extremes of expression.

A shy, withdrawn child, for example, would benefit from the follow-the-leader games described earlier in this chapter. Parading around the house in your wake, copying your movements and other musical activities can help a timid child experience a larger body sense and explore sounds and movements that he probably wouldn't have made otherwise. Making a variety of sounds—high and low pitches, open and closed vowel sounds—while you march encourages him to vocalize and use his voice. When it's his turn to lead, he feels empowered. He has something to teach you for a change, and his self-confidence receives a boost from the idea that you are interested enough to follow and echo him.

If your child seems to have a somewhat fearful or extremely sensitive nature, create a collection of half a dozen or so lullaby tapes and play them for him over and over. He will appreciate the consistency of the music, and as he listens to it and learns the words, his confidence may increase a bit. Such children are also frequently comforted by personalized songs their parents invent about their own life. You might create some of these, featuring your child, his friends, pets, family members, and so on. Sing the songs while holding and rocking him, and record them so he can play them in bed.

Aggressiveness can be quite a problem at this age, whether or not it is an integral part of your child's inherited temperament. After all, this is a time of experimental biting, hitting, and frequent yelling of "No!" and "Mine!" If your child frequently acts out, play some light pop music as soon as you see him beginning to reach his limit. Research has shown that such music reduces the rate of inappropriate or disruptive behavior among preschoolers and older children. Playing a recording of calming rhythms quietly in the background, such as the Minuet from Leopold Mozart's *Toy Symphony*, might also serve to calm your child down.

Overexcitement and moving at too fast a pace can also be a challenge for children not yet able to manage their own energy level. During my days teaching music to young children, I frequently encountered children who were too wound up to be able to sit still and hear what I was attempting to teach them. I found that by sitting such a child on my lap with his back to me and holding a small drum gently against his stomach, I could slow his rhythms down. "Here, sing a song with me," I might say, beating the drum slowly and lightly as I sang a familiar tune. The outer rhythm of the music created a kind of rhythmic organization in the body of the child, and the change in the way he participated in the rest of the class was practically immediate.

In general, research has shown that music has a powerful effect not only on mood but also on perception and attitude. Amazingly enough, what children and adults, for that matter, retain from a particular experience can depend on the tone of voice used or even on what music is playing in the background. Words are better remembered as being positive if bright, light music is playing, and words are considered to be negative when accompanied by slower, heavier sound. The type of background music affects how people characterize paintings, but the paintings' mood doesn't alter the way we perceive music. After listening to depressing music, experimental subjects judge neutral faces as expressing more rejection or sadness than invitation or happiness, even though such emotions weren't actually present in those faces. As is obvious to moviegoers, the mood of whatever music happens to be playing greatly affects both what people feel at the moment and recall later.

All of which goes to show how important it is to attend to what music your child is listening, playing, and resting to. Take some time to really listen to the music that fills your young child's days. Does it challenge his intellect while offering supportive emotional content that will help him build self-esteem? An undiluted diet of popular mass-market melodies may be what a toddler asks for, but it won't nourish the mind and body as well as a varied menu that includes classical music and traditional children's songs. A child's obsession with one mediocre musical selection may be a sign that he hasn't been exposed to enough high-quality recordings or live music performances. Keep in mind as you turn on the television or load a stack of CDs that music is not just songs that we teach our children to sing. It's a key we give them to unlock their minds, bodies, and hearts.

A SOUND SOLUTION

SSSSH!

By the latter half of the second year, your toddler is able to begin managing his movements and noise level to a certain degree. You can help curb the desire to screech and scream by repeating in a whisper what your child has screamed. Chances are your child will stop, stare at you in great surprise, and begin to scream again. Interrupt him, mimicking his noise in a slower, softer whisper. You might even mime his actions, which is sure to provoke a giggle. In short, he won't be able to resist the game and soon will be happily whispering, too.

THE DIAPER DANCE

"Each week when the babies arrive for their class, Mozart's *Eine Kleine Nachtmusik* is playing softly in the background," music educator Dorothy Jones tells me. "Many families have chosen this piece for listening at home as well. One family told us that as their son began to speak he would stand in front of the stereo and demand 'my Moz.' After he started walking, this same child did not like to have his diaper changed. His mom told him she would turn on his Mozart while she changed the diaper. Every time he would stop what he was doing and lie down immediately."

Clearly, the mother in this story learned quite well how to use entrainment, the act of inviting a child to partake in the mood of a particular piece of music, to make her child's life and her own much easier. During the toddler years, entrainment techniques continue to help your child work through the emotional challenges he will inevitably experience at this age. One of the most difficult aspects of childhood for a toddler, however, is the act of transition—moving from playtime to nap time, for example, or moving from pajamas to clothes! Transitions, which by nature occur over a period of time, are not as easily smoothed over with one musical mood or style. In these cases, it might be best to use another musical technique called the *iso principle.*

The iso principle involves a gradual change of pace in rhythm, speech, or emotional content that brings a person from one physical or emotional state into another. Ravel's *Bolero* exemplifies this principle, as does the "Happy Birthday" song, with its joyous, celebratory beginning and slower, more thoughtful completion. It is easy to see how music can be used to express understanding of a toddler's misery, for example, and then gradually bring the child to a healthier, more balanced and relaxed state. Let's say your child is moving perilously close to a tantrum because you bought him an ice cream cone yesterday but refused to do so today. To pull him out of it, you might match the tone of his anger without, of course, matching the anger itself, then gradually turn those sounds into happier ones, just as you did with his infant cries.

The iso principle can be used to transition from playtime to nap time, sleep to wakefulness, a quiet book to a lively dance around the coffee table. It might even be fun, on some weekend day when you're hanging around the house, to experiment with creating a Mozartian musical score to accompany your child through all the activities of a typical day. Go through your Mozart recordings, choosing lively selections such as the *Toy Symphony,* minuets for times when your child is actively playing; less intrusive pieces such as the Andante movement of Symphony No. 6 (K. 43) while he's coloring or working at his table, soothing selections from Mozart's many adagios during rest time, and transitional music such as the Flute Quartet in C Major (K. 171) for the periods in between. Or, if you like, broaden your selections to include other kinds of music. A perfect sound diet for a rainy day might consist of Mozart and other romantic and baroque classics for about 30 percent of the day; children's songs and lively pop tunes for another 30 percent; and songs and music from other cultures for about 10 percent. During the remaining time, you can all enjoy a little silence.

It is fascinating to see how effective music can be in turning a typically chaotic routine into a relatively serene sequence of moods. Just remember: the basic concept involves starting at the rhythm or mood where your child presently is, then gradually speeding or slowing the tempo, modifying the emotional content, enthusiasm, and melodic variety of the music, to help your child get from here to there with ease.

SINGING IN ONE VOICE

"Duncan came with his mother to meet me before we enrolled him in the Sunrise Music program," wrote Jean Schoenfelder, a teacher in the Music for Young Children program in British Columbia, Canada. "His mother had some concerns about his ability to participate . . . and with good cause. When I met him, he would not look at me and heaven forbid that I should speak to him. He would tell the wall that he was looking at the wall just so I would get the message that he wasn't interested in my attention. His mom felt that getting him involved in a music program might draw him out because he apparently was quite a musician for a two-and-a-half-year-old at home. We signed him up then.

"When his lessons began, they were always the first to arrive each week, but Duncan was as antisocial as I have ever seen anyone in my life. For the first ten-week block, he never spoke or sang in class, never got up to do actions or dancing, never responded to anyone there except his mother. He did like the rhythm instruments and would play any instrument I gave him, so I tried to make sure we were heavy on the instruments and within reason, as there were eleven other kids who did like to sing and dance.

"Block Two started just after Christmas, and Duncan began much the same, but pretty soon we noticed some changes. He would speak to me before or after classes now, and would clap along to the tunes. He still loved the instrument portion and the coloring. He was listening in class. At home he apparently sang all the songs with much enthusiasm and volume, so his mother was comfortable with bringing him to a third block. She had expressed her wish to put him into preschool but still felt that he wasn't going to be ready in the fall.

"Block Three was like a miracle had occurred. Duncan started to sing occasionally in class. He would get up all the time and do the actions, and most surprisingly he began to call out the correct answers to questions I gave them. I had planned to include the Sunrise class in my May recital, and the children worked hard on 'I Had a Bird.'

"The recital day came and all the kids performed well, but I really expected Duncan to not sing, because we had easily 150 people in the room. When the singing started, though, he sang, too. He sang the whole song, made all the appropriate sounds, and bowed at the end. I could have cried on the spot. After the recital he came up, looked me in the eye, and said he looked forward to music lessons in September. His mom added that he would be attending preschool after all."

Nothing brings children into one voice like music. As your child moves steadily between ages two and three from contented isolation to equally contented parallel play, and finally the beginnings of emotionally fraught play dates and true friendship, music is there to teach him the social skills of turn taking, listening, and sharing. A social life is vital at this age not only for pleasure's sake but also, interestingly, because children learn best during this period by imitating other children. Group music activity has been shown, as we saw with Duncan, to improve young children's social skills. An early childhood music class is bound to fill your young one with the sense of community and belonging that he craves, especially now that he's old enough to join in the singing.

At least as important are your child's musical connections to his family members, particularly his grandparents and the other senior members of his clan. As your child's musical culture fades in the face of today's prepackaged passive music experiences, his grandparents may be the only ones who remember the body-, brain-, and emotion-building songs and games of decades ago. Child psychiatrist Arthur Kornhaber tells us that "the attachment between grandparent and grandchild is second in emotional power and influence only to the relationship between parents and children." What's more, they can remember back further. Encourage them to pass their memories on to their twenty-first-century grandchild. And don't forget to ask your child's caregiver for contributions from his or her own generation, hometown, and culture. It is amazing how much music there is to share with today's children if we only open up our hearts and sing.

A Mozart Musical Menu

- Adagio from the Divertimento (K. 287). The Italian word *adagio* means to put at ease. This music is slow and leisurely, perfect for an afternoon nap. Invite your child to close his eyes and let his body rest as Mozart takes him on a relaxed journey through sound.
- March No. 1 (K. 335). Here is a fine march for preparing your child to get dressed and go outside. Its lively beat will put mind and body in motion and in order, and inspire your little one to move.
- "Champagne Aria" from *Don Giovanni.* Dance. Skip. Twist. Turn. Invent wild motions. Dance along with this delightful aria. You can dance along with your child or use this for your own aerobic exercise.

CHAPTER 6

TAP, RAP, AND SING ALONG

The Seeds of Creativity (Three to Four Years)

I am happier when I have something to compose,
for that, after all, is my sole delight and passion.
—WOLFGANG AMADEUS MOZART

As a young child, Mozart swam in a sea of music. His Salzburg home resonated with the lilting tones of keyboard melodies, violin sonatas, and chamber music, charming pedestrians on the street below. At age three or four, the budding prodigy had already begun participating in rehearsals with his father, mother, and elder sister, creating his own first melodies based on the musical and emotional knowledge he had absorbed. Bone-deep familiarity with pure tone and rhythm, constantly repeated and playfully varied, gave young Mozart the confidence he needed to take risks and experiment in his own creative work. Gifted performer by age four, composer of his first minuet at age five, Mozart had demonstrated his abilities on the clavier for the empress of Austria by the time he was six years old. From the beginning, his sprightly, deceptively transparent style captivated his listeners. It was a style whose celebration of human creativity was rooted in those earliest years of musical pleasure.

Of course, a prodigy at the level of Mozart is a rare occurrence indeed, but the fact remains that any child's delight in creative pursuits, as well as her ability to *think* creatively, can be encouraged through stimulating encounters with sound. Your three-year-old has already learned the ways in which rhythm and tone help her move with greater efficiency and grace, and has used music to explore the emotional subtext of her widening world. Now, as she enters childhood, it will seem only natural for her to *respond* to the music inside her, to take the leap toward inventing her own songs, games, stories, and musical and nonmusical imaginary play. Listening to the music of the masters, she will begin to express her own ideas and feelings in dance and song. By widening her musical repertoire and making up songs with you, she will expand her vocabulary and enhance her verbal skills. Increased experience in movement will add to her new understanding of time, space, basic mathematics, and other abstract concepts. Meanwhile, her deepening relationship with music, fortified, perhaps, by her first opportunities to watch professional musicians at work, may lead her to begin learning an instrument herself.

Call and Response

"Will and I were taking the bus home from preschool the other day," a musician friend told me recently. "He jabbered away from the moment we boarded the bus, to me, to the passengers sitting behind us, and to the people standing in the aisle. He completely monopolized one woman standing nearby. I guess she had a friendly look that made him think she would listen. 'Hi! I'm Will!' he said. 'I was at school today! We made flowers out of paper! I cut with scissors! I didn't hurt myself!' And on and on and on. After about ten minutes of this nonstop monologue, the woman leaned close to him and said in a confidential voice, 'You're four, aren't you?' Everyone around us laughed as Will said, very surprised, 'Yes, I am!' "

It's true, healthy four-year-olds are very easily identifiable by their exuberance, love of human interaction, and overwhelming desire to give back to the world some of the energy they receive. Within this desire to respond lie the seeds of mature creativity. The more you encourage and support your child's experiments in creating her own, original voice, the better equipped she will be later to think creatively, solve complex problems, rely on her own sense of correctness rather than on others' opinions, and find pleasure in the machinations of her singular mind. Much of this work can be accomplished within the context of music, movement, and sound.

Part of the reason for this leap in creative activity is the continued maturation of your child's brain. By age three or four, the basic neuronal systems are plugged in and working, and the neural net has first expanded and then been pruned. As links between the various areas of the brain continue to be insulated and reinforced, they begin to work together as a much greater, more dazzling whole. At around age three, for example, strong connecting pathways are laid down in the brain's rear associative network, a large, mostly unspecialized area covering nearly the entire back half of the head. These pathways strengthen the links between the auditory and visual centers, the visual and motor areas, and the auditory and motor areas, leading to much greater coordination among hearing, seeing, and action. As a result, your child will be able for the first time to stop, start, and change direction suddenly; to imitate your clapping or stamping with greater accuracy; to slide, whirl, hop, gallop, and walk on tiptoe; and to play a percussion instrument with apparent ease.

Your little one's mental ability will undergo a process of refinement and enhancement during this period as well. Increasingly, she will be able to think symbolically, in terms of abstract concepts. She will process information more efficiently. Her attention span will lengthen. Both her ability to reason logically and to think intuitively will be enhanced. She will be able to consider more than one mental dimension simultaneously, such as the size and color of an object. She will also be able to more accurately re-create past events, thoughts, and emotional experiences in her mind, and sometimes relate them to you.

All of these skills clearly help your preschooler interact more fully with the outside world as it floods her senses with all its harmony, noise, and splendor. The honking of car horns, the barking of dogs, the roar of dump trucks, and the intricacies of language all command your child's intense, if still relatively fleeting, attention. Sounds fill her dreams, her fantasies, her every conscious moment. Her ears are ecstatic with the beauty and variety of vibration. As she listens, contemplates, and categorizes, she is increasingly able to compare what she hears to her existing store of knowledge, and use her expanding language skills to create a unique reflection of her experience. Now she can and will judge whether an act is "good" or "bad," a person is "nice" or "mean," or a musical selection is "funny" or "sad." She can respond to an angry word with several of her own and she can listen to Mozart and invent a dance to express how joyful the music makes her feel.

TUNE IN, TUNE UP

MUSICAL EYES

Select a favorite Mozart piece to listen to with your child. (The Catillion and the Allegro from the Divertimento in B Flat [K. 15] is a good choice.) While listening, discuss with her what pictures, colors, or stories the music brings to mind for each of you. Then play the selection again, encouraging her to act out these images with her body—impersonating, say, a happy butterfly when listening to one section, or a busy little groundhog when listening to another. The more specific you are with your suggestions (an excited, twirling dancing princess rather than just a dancing princess, for example), the more you will stimulate her imagination. Let her suggest some ways for you to act out the music, too. Not only will this interlude exercise your child's mind, but it may well create some happy memories for you both in the years to come.

CREATING A VOICE

"I am a full-time mom, passionate about giving my two marvelous and precocious sons, aged two and four years, every possible advantage in life," writes Debi Anderson-Wilde of West Valley, Utah. "I read about your work relating to the Mozart Effect and thought I would give the idea of learning enhancement through Mozart a try. I put on some music from my own collection as my older child and I had our afternoon thinking and learning time. Suddenly and startlingly, he began to SING all of his responses to me! This was new and completely different! I'm certainly convinced that something was stimulated, and I'm enthused about finding out just what it was."

Certainly, the connection between music and thought are very strong at this age, particularly if the child has been surrounded by music from her earliest days. As children begin to express themselves, they frequently pick up the sounds in their environment and put them together in new ways, building a primitive structure for their thoughts. Music's natural rhythms, expressive melodies, and sensory pleasures provide a child with perfect prefabricated units of expression that mirror what they want to say. For this reason, a child of three or four is almost as likely to sing her thoughts as to say them. As one mother told me, "I was gardening the other day, trimming a honeysuckle vine, while my three-year-old wandered around the yard singing to herself. I ignored her song until she passed close by and I realized that she was singing, ". . . why are you cutting the vine . . . ?"

The link between music and self-expression in your child is instinctive, and it runs very deep. Think back to those evenings when you played musical recordings for your two-year-old while you made dinner. Remember how emotionally involved she became in the music, sprawling on the kitchen floor, listening raptly, as though she'd forgotten her body in her effort to feel the music? Back then, her chin may have quivered and her eyes filled with tears over a turn in a musical phrase, or she may have suddenly jumped to her feet with joy. Now, though, feeling the music is probably not enough for her. She must sing along with and/or dance to the song, continue singing it or make up her own song after it's finished, tell you whether or not she likes the selection, or let it spur her to tell you a long, involved story about her Raggedy Ann doll's visit to the playground today. In other words, she must use the music's energy to create something new.

As always, the way to encourage and enhance this growth is to start with what your child already knows and loves and expand from there. Certainly, by now, your child loves sound. The melodies and rhythms of childhood have no doubt slipped under her skin, and this sensory, physical, and emotional connection offers a conduit from the unconscious to conscious expression. You have already helped her increase her awareness of the musical qualities of familiar objects, such as the squeak of a door, the ring of the telephone, or the click-click of the dog's toenails on the floor. By asking such questions now as, "That telephone has a high sound. What else makes a high sound? Can you make that sound?" you encourage a more complex, creative awareness of her world.

In the same way, moving from the simple pleasures of singing a song toward actively thinking about it can stimulate the beginnings of creative thought. You might, for example, sing a familiar nursery song to your child, then sing the song again, leaving out the last word of every other line and letting her fill them in. For example: "This old man, he played one, he played knick-knack on his ____." This is a simple challenge for two- and three-years-olds and will leave them giggling triumphantly when they find they can meet it.

Once your child has grown familiar with the pleasures of filling in the blank, try singing a song that she doesn't already know. Before the last line of the song, stop singing. Ask your child how she would complete the song. Let her sing the ending to you. Then respond with the real ending of the song. Talk about the differences between her ending and the actual one, and decide which one the two of you like best. From here, it is a short step toward inventing your own songs about people, objects, and activities in your lives, a fun activity that will stimulate her awareness and thinking about those subjects and encourage her love of language.

Of course, movement can be as creative an act as inventing new lyrics to an old melody. Young children love to dance, hop, and sway to recorded and live music. As you now play your little one's favorite selections, ask her what the various sections of the pieces remind her of. Then encourage her to express these ideas through movement by tiptoeing to the soft sections and stomping her feet to the loud parts—or moving in slow motion with Mozart's Adagio from the *Gran Partita*, and dancing more rapidly to the *Rondo à la Turke*. Show her how to raise her hand in response to rising pitches and lower it as the pitch descends or to raise and lower her whole body as the music modulates. It can also be fun to sing a song that your child knows, singing all the action words in a soft voice as she acts them out. ("Oh, Susannah, now <u>don't</u> you cry for me/'cause I <u>come</u> from Alabama with a banjo on my knee.") After you've finished the song the first time through, sing it again, this time singing all the emotional words or phrases softly while your little one acts them out. ("Oh, Susannah, now don't you cry for me/'cause I come from Alabama with a banjo on my knee."). Next, sing all the nouns and let your child act out those words. Finally (if she isn't too tired) sing the song again, slowly enough to allow your child to act out each word or phrase she's acted out before. In this way, she learns not only to listen but also to use her body, mind, and spirit to express and create.

Music making naturally evolves from moving to and expressing the music one hears. Now, more than ever, it's a good idea to keep a variety of child-friendly musical instruments around the house, such as a small drum, egg-shaped shakers, jingle bells, and perhaps some kind of keyboard. It's also fun to show your child how to make music with her own body when other instruments aren't around. Studies have demonstrated that three- to five-year-olds naturally and freely experiment with musical instruments when they're available, creating songs and imitating rhythms with their bodily movements. These songs have been shown to have very clear organizational patterns, with a restricted range of pitch intervals but with distinct brief melodies. In other words, your child is not just banging on that keyboard, no matter how it sounds to you; she is practicing very deliberately with pattern and form. For this reason, it is usually best to leave her to her own experiments, rather than trying to make her play an instrument with instant correctness.

You can, though, provide her with as wide a variety of musical models as possible by playing recorded melodies from her own culture and around the world, and by singing new songs with her frequently. You can also give her new ideas for creative ways to play with music. Next time you tell or read her a familiar story, let her add the sound effects. If a character "runs away home," your child might shake a set of bells to simulate the sound of running. If an acorn falls on Henny Penny's head, rhythm sticks can make the appropriate sound at the right time. In stories with their own chants (such as "Run, run, as fast as you can, you can't catch me, I'm the Gingerbread Man!") she can try to keep the beat of the chant with maracas or hand clapping though she probably won't be able to succeed completely until age four. She might even enjoy establishing a mood with keyboard or xylophone music. The more your child is allowed to experiment informally with music in this way, the more comfortable she will become with the act of throwing herself un-self-consciously into creative activity. Clearly, this is a much more vital skill than the ability to play a violin accurately, sing a song on pitch, or perform a perfect pirouette. And best of all, you don't have to be a trained musician to participate!

A MUSICAL RECIPE

CHANT ALONG, MAKE UP A SONG

During this period, you are bound to overhear your child singing her thoughts out loud—perhaps unconsciously. Usually, the song consists of a familiar melody with the child's own lyrics slipped in. Frequently, part of one melody will be combined with part of another; and your child may add an additional original flourish or two. Whatever the result, take advantage of this opportunity to point out to your child that she has done something creative and fun. Ask her to repeat the song, and sing it along with her. If you are able, write the song out in musical notation on a piece of paper. If you don't know how to write music, just draw a short line for each note, higher or lower on the page to indicate the relative pitch. Then write the lyrics underneath, as with this creation (sung to the tune of "Ring around the Rosey"):

I ate my ice cream

Then illustrate the song with a little drawing, just as in a children's songbook. Sing the song again, perhaps clapping along, to show your child how much you enjoy it. Later, sing it to other family members and encourage them, outside of your child's hearing, to casually hum or sing the song to themselves in her presence, so she can have the experience of watching her creation become a part of the family culture. Watching something she created naturally and unthinkingly become a work of real value is a powerful experience for a very young child. She will feel validated by your interest in her creative play and will focus more consciously on those skills. As you continue to write down and enjoy her songs over the months that follow without overdoing it, which may make her feel self-conscious, she will come to think of herself as a creative person whose thoughts and ideas are important.

As your child grows and develops, she will begin to choose her own forms of creativity, musical and otherwise. Deciding that she'd like to invent a song about her walk through the park, and then doing it, will give her a satisfaction that guarantees more creative efforts in the future. Gradually, she will find that her songs, dances, chants, and other creative activities can serve as an emotional outlet, helping her to give solace and encouragement to herself when no one is there. She will be grateful for your pride in her efforts but will begin to evaluate her own work, too. Soon, she will be able to say to herself with satisfaction, "I stuck with it," or "I thought of a new way and it worked."

ARITHMETIC-TOCK: THE RHYTHMS OF THOUGHT

As your child's attention span, communication skills, and ability to think abstractly improve during this period, it is sometimes easy to forget that she still doesn't really think like you. Until about age eight, children continue to experience the world mostly through their bodies—that is, a discussion about numbers had better include objects to hold or look at, or a steady beat. On the other hand, your three- to four-year-old has moved past her ability to just recite numbers and letters by rote to her first true understanding of what they mean. This means that you can begin to introduce the basics of such symbol-dependent subjects as mathematical problem solving and storytelling, as long as you ground your discussion in concrete experience.

In fact, children don't consider math an abstract concept at all, and in a certain sense they're right. For them, mathematics consists of the properties of objects and their relationships to one another, qualities that your child explores concretely every day. Separating objects by shape or color, counting them, deciding who has more and who has less, pouring sand from one container to another, chanting nursery rhymes, and singing "Ten Little Fingers" are all ways in which math is understood by a three- or four-year-old. You can encourage her to expand on this knowledge by helping her count the number of times she can bounce a ball, the number of seconds she can balance on one foot, or the number of ways you let her move her head. Once she understands the *concept* rather than just the rote sequence of numbers, she will be able to count backward, too. You can help her learn how by helping her pretend to be a rocket ship and announcing the countdown, letting her rise up a little with each number until she "blasts off" by leaping into the air. "There Were Ten in the Bed and the Little One Said . . ." is another physical way to introduce the same mathematical concept. All the more fun if it's acted out.

Mathematics is not the only concept your child can explore most effectively with movement and song. Language is another skill ripe for development, and your child is no doubt actively seeking ideas. She soaks up words like a sponge at this age, acquiring a new word every two and a half hours, according to one study, and can't wait to put her vocabulary to use in long, involved conversations with anyone who will listen. Obviously, this is a window of opportunity not to be ignored and, again, the key to stimulating her language acquisition is to make the experience pleasurable. Read aloud and sing to her every single day. Teach her as many songs involving wordplay and new words as you can find, such as "This Old Man," "Michael Finnigan," and "Willoughby Wallaby Woo." Help her invent her own chants, tongue twisters, and rhymes. To improve the quality of her expression, talk with her about how different objects look, sound, taste, and feel. Encourage her to talk about her own day-to-day experiences and create songs about them. Have her depict these experiences through movement and then discuss with you what she was doing. It doesn't matter whether her songs are in the form of simple chants, elegant variations on nursery rhymes, or creative rap. All forms can contribute vitally to language development.

Other rhythmic and musical games can supplement the skills that underlie the ability to tell a good story. Sequence songs such as "The Hokey Pokey" and "Head and Shoulders, Knees and Toes" stimulate your child's sequential memory. You can begin by describing your actions as you perform them, then eliminate the actions, and lengthen the sequences as she's ready. You might also take advantage of her longer attention span to play recorded music that tells stories, such as the overture to *William Tell* by Rossini. By helping your child act the stories out with puppets or her own body while listening to the music, you let her begin to get a sense of a story's natural shape.

Of course, no one parent has the time or energy to engage in every one of these activities and neither does your child. As always, it's more important to have fun with your child than to push her, or to try to cram every activity into your time with her. In general, it's best to keep things light and simple. Don't introduce too many new games, words, or concepts at a time. And always follow your child's lead. When she begins to look tired or bored, put the music away.

Of course, education at this age has little to do with imparting facts, monitoring progress, and expecting concrete results. Your goal and your child's is rather to prepare a solid emotional, physical, and neurological grounding for the more formal schooling ahead. Through music, you can continue to listen to your child, help her make sense of her world, and gradually guide her toward her own true voice. By exposing her to high-quality classical music and children's songs, you can help build the neurological tracks that will serve her well as a thinking, self-directing, independent adult. Taking her along to parades, live music performances at bookstores or public areas, and children's musical events will not only introduce her to the musical culture of her community, but will also allow her to witness the beauty of music as it is being created.

Music, with its complex combination of rhythm, pitch, and pattern, mirrors the kind of multidimensional neurological activity that leads to creative thinking. As Dr. Dee Coulter puts it, "Music has a fundamental integrity or honesty or truth or beauty about it, and one listens to it and recognizes that it is so."

FOR YOUR CHILD'S HEALTH

SONIC SPEED LIMIT

As children enter their first preschool class or playgroup, many parents find that the adults in charge assess them in quite different, and often surprising, ways. One of the most common assessments given a child of this age is that he or she is hyperactive. Of course, hyperactivity is not a precise term, and these days it can often simply be a result of too much auditory or visual stimulation or stress in the home or elsewhere. If your home typically contains a noisy television, computer games, videos, radio, and the noise of other children bouncing a ball or shouting at one another, your preschooler is most likely overstimulated. Turn down the volume at home, creating a predictable routine that this sensitive child can rely on. Too much sound in an environment makes any child speedy. Even presenting meals at the same time every day, managing parental stress, keeping the house neat and orderly, and providing a separate room in which she can find peace and quiet can often help dispel the symptoms of restlessness and anxiety that called attention to her at preschool.

Providing a child-size set of drums and other musical instruments with which to release stress may also help. Join her in music making, drumming, bell playing, or dancing games that start out fast and gradually, over the space of seven or eight minutes, slow down in pace. You might put a recording of soothing rhythms or light jazz music on the stereo, keeping it turned low, or even set a metronome on a slow beat (fifty-eight to seventy-two beats per minute) and sit with your child, holding her hands and softly counting out the ticks. The goal is to help your child learn to help herself at those times when she feels overstimulated in order to help her organize the world, the space, and the timing around her in a way that will make her environment feel more safe and reliable.

INSTRUMENTS OF PLAY

When young Mozart was four years old, he tried to compose his first piano concerto. Papa Leopold and a friend discovered him writing out a smudged but legible score. When they commented that it was hard to perform, Mozart played it, declaring, "That's why it's a concerto. You must practice hard to play a concerto."

This possibly apocryphal story speaks to a number of expectations surrounding the idea of music instruction for preschoolers: the desire to instill rigor, self-discipline, musical enjoyment, and even a sense of culture in a young child. But another aspect of the story should not go overlooked: that is, young Wolfgang was composing his concerto *voluntarily,* for the sheer joy of it, without parental pressure.

Certainly, the three- and four-year-old's greater mastery of fine motor skills and hand-eye coordination and her ability to stay focused on a task for a slightly longer period of time can make this period an excellent window of opportunity for beginning instruction in an instrument. Introducing your child to instrument training is beneficial not only for musicality's sake but also because the experience of keyboard instruction, group singing, etc., has been linked in many recent studies with an increase in spatial and temporal abilities (abstract thinking skills that facilitate higher-order learning); greater creativity; improved math performance, aesthetic awareness, social interaction, and academic skills; and even an enhanced ability to learn a foreign language! Anecdotes also abound about its ability to help children with sensory integration problems.

One study at McGill University in Montreal followed a group of economically disadvantaged children over a two- to three-year period. Half the students were given free piano and keyboard lessons; the other half were not. The children who received instrumental instruction went on to perform at the top of their class academically, while the control group stayed the same. In another study, first-graders who were given thirty minutes of daily music instruction for a year exhibited significant increases in both creativity and perceptual-motor skills, compared to an equivalent class that did not receive such instruction. Adults who received music training as children have demonstrated more coherence between the hemispheres of the brain than those who did not, and have a better memory for words.

Research indicates that actively making music has a substantially greater beneficial effect than listening to it. As violin instructor Joanne Bath puts it, "I have taught more than two hundred Suzuki violin students privately, in a thirty-two-year career in Greenville, North Carolina. They all listen to music for hours each day, starting at the age of three or earlier. They *all* are outstanding academic students, as well as being fine violinists."

There is absolutely no rush to enroll your child in instrumental instruction unless she shows an interest in learning to play. Certainly, the enjoyment of music making should be the primary reason for her to begin to study an instrument. In fact, though, many three- and four-year-olds do like to take lessons, especially when it involves making music with an instructor or with other children. If your child does show a natural interest in starting music instruction, there are a number of instruments, and a variety of instructional methods, to choose from. Many young children are drawn to the visual logic of the piano keyboard; others greatly enjoy the warm physical vibrations of the violin. Before you make your choice, visit a music class or attend an early-childhood recital with your child. Let her watch and listen to other children perform. Then follow her intuition regarding which instrument she would like to play.

SPOTLIGHT ON THE SPECIALIST

SHINICHI SUZUKI

The son of a violin maker, Shinichi Suzuki learned to play the instrument as a child and, after studying in Berlin in the 1920s, began teaching music in Japan. Dr. Suzuki based his method on how easily and naturally children learn to speak their mother tongue, no matter how complex, through constant repetition and encouragement from their parents. He theorized that music could be taught in the same way. In learning language, a child speaks first and then reads words. Suzuki therefore excluded most discussion of music theory from his classrooms, insisting that students must first play well, and only then start to read music. In language development, a child hears and says words again and again. Similarly, Dr. Suzuki believed a child should hear and play his music repertoire repeatedly, learning both technique and music in small, sequential steps. As a result, Suzuki practice sessions are supplemented by listening to recorded versions of the same music daily.

Suzuki also insisted on the importance of parental involvement. In Suzuki classes, parents participate in a Suzuki Triangle of parent (or other adult), teacher, and student—the parent learns the basics, observes the teacher and child, and acts as the child's violin teacher at home. Children participate in group lessons as well as individual instruction, so that they can learn from one another and enjoy the learning process.

Suzuki emphasized his belief that every child has unlimited potential when she is nurtured by love. Today, more than 300,000 students and teachers use the Suzuki philosophy and method, which has spread to countries all over the world and is used to teach violin, viola, cello, bass, flute, piano, guitar, recorder, and harp. It has also been incorporated into a number of curricula in preschools, elementary schools, and other early childhood programs. Special needs children have particularly enjoyed the benefits of learning in a supportive group setting. The lives of autistic, blind, and deaf children, as well as those with Down's syndrome, have all been greatly enhanced by the flexible interactivity of the Suzuki Method, and some have gone on to perform admirably at an advanced level.

The most important aspect of music instruction for your child at this age is *your* involvement. Your participation will not only improve your child's music performance, but will also increase her pleasure in the process and thereby her motivation. It also ensures that the beneficial side effects of music instruction will take hold. A 1998 study at Sam Houston State University demonstrated that, while early music training enhances children's intelligence leading to improved scores on abstract-reasoning IQ tests, parental involvement in the training greatly affects the amount of improvement. In fact, the scores of the children studied improved in direct proportion to the level of the parent-child participation.

Make the effort to establish a musical relationship with your child as her instruction begins. Attend lessons with her, learn the basics of the instrument yourself, and invite her to share her observations regarding the process as she proceeds through the program. As with the early-childhood music programs discussed earlier, the individual instructor is a vital element in the learning process. Make sure that your child and a potential teacher are compatible, and that the teacher not only loves working with kids but is well informed on how to do so. Teaching the love of music is as important as teaching musical skills.

Some children, especially those with learning disabilities or physical impairments, may become frustrated with the lessons. If this is the case with your child, help get her back on track by breaking each task into very small steps, and don't insist that every step be performed perfectly every time; allow her to take breaks; acknowledge how hard certain challenges can be; praise her for the progress she's made; and take vacations from practicing once in a while. Sometimes, joking around about the frustration, such as drawing a silly picture of a frustrated performer, is enough to break the tension. In the end, the effort to master an instrument is usually worth the occasional frustration for both parents and child. As your little one progresses from listening to "Twinkle Twinkle, Little Star" on her Suzuki tape to playing it on her piano, she will gain in self-confidence and self-esteem. Besides, as one mother said, "Isn't it a thrill when a very young child can sing the Bach minuets after hearing them on the first Suzuki record?"

Feeding the Ears and Heart

"One of the most exciting things about becoming involved in a good Suzuki-type program is the opportunity to become part of a real community of parents and children who provide a stimulating and encouraging environment in which your child will thrive," Elizabeth Mills, coauthor of the book *The Suzuki Concept: An Introduction to a Successful Method for Early Music Education,* wrote to a friend who asked for advice in choosing a music program for her child. "Not only your child will develop, you and your husband will find help in growing as parents from the others who have been in the program long enough to help you over the hurdles."

As your child grows and is able to take part in more and more communal experiences, you will increasingly experience the joys of integrating music into your community life. Seasonal musical events such as Christmas caroling, summer concerts in the park, and parades down Main Street on the Fourth of July will fill her with a wonderful sense of security and well-being of being a part of a big group, part of the larger pattern. Alice Cash, a professional musician and social worker who was in the middle of analyzing the music of Beethoven on a National Endowment for the Arts grant when she learned she was pregnant with daughter Julia, began taking Julia to free concerts when she was two. "We'd sit at the back of the auditorium with crayons, paper, and M&M's," she wrote to me. "From the beginning she was attentive, although a few times we had to slip out. I would recommend taking children early to live concerts, but sit near the back, as it's not fair to the performers to have fussy children in the audience."

Julia grew up to become a professional violinist herself. Even for those not destined to play music professionally, experiencing music in community is not only an exciting experience, but it also underscores the wonderful powers of empathy and creative expression that music offers to us all.

A Mozart Musical Menu

- Concertante from the Serenade No. 9 (K. 320). Mozart composed serenades for gala events in royal palaces. This music was used to create a festive atmosphere at parties and for meeting people. Help your child imagine a fine procession of people dressed up in their best for a great party in the wigs, great hoop skirts, and fancy attire that Mozart must have seen as she listens to this wonderful concertante.
- *Rondo à la Turke.* This famous rondo is a great piece for clapping along, moving to the music, and listening for the accents in the rhythm. A great workout for mind and body.
- The Catillion and Allegro from the Divertimento in B Flat (K. 15). This is a perfect hide-and-seek piece. Look around, find a friend, run and hide to the music, and then sit down and play it again while you rest.
- "Papageno's Song" from *The Magic Flute* (K. 620). Help your child imagine playing a magic flute as she skips, dances, and twirls, bringing the magic of Mozart's beauty to you and to the ears of her friends.

CHAPTER 7

SING . . . SING A SONG

Social Interaction (Four to Six Years)

When I hear music, I fear no danger. I am invulnerable. I see no foe. I am related to the earliest times and to the latest.
—HENRY DAVID THOREAU

When Mozart was only five, he and his sister were taken to Vienna to perform at the palace of the Empress Maria Theresa. Mozart slipped and fell on the palace's marble floor and was helped to his feet by six-year-old Princess Marie Antoinette. Seizing the moment, Mozart proposed marriage to the princess. The empress was so charmed by his social acumen that she presented Mozart and his sister with beautiful matching sets of clothes.

Your child may not be meeting any princesses at his local kindergarten or grade school, but you can be sure that he's becoming increasingly interested in getting to know his peers. Adults may focus on their children's cognitive skills as they enter school, but the kids are quite naturally and rightly focused on the new social opportunities. Having only recently learned to capture and hold another person's attention, your child is giddy with the prospect of making new friends with whom to carry on intense conversations. This is the age when he will spend hours at a time building block towers with one or two other children, or scream with joy and wildly embrace a best friend at the beginning of every single school day. Learning to master the intricate dance of social interaction, the subtleties of body language, personal space, verbal rhythms, and sharing of the spotlight, will bring him a great deal of satisfaction in the years to come. Music is an excellent tool for teaching the first few simple steps of this very important aspect of growing up.

Fortunately, your child has already learned to look to music for personal comfort, joy, and stimulation of mind and body. Now, more than ever, rhythm and tone can help him fill in the social and emotional blanks that sometimes occur in our schools as children's heads are attended to but their hearts are neglected. It can help bolster his confidence as he faces a large classroom and intimidating peers and teachers for the very first time. By paying attention to sound and its effects, you can discover whether his hearing or his classroom's noise level is interfering with his learning. His favorite music can even help him weather a physical challenge if he has an accident or falls ill. In short, once again, as your child walks through that door to begin his first day of school, music can be a friend to support and guide him, deepening his experience and enhancing his daily life.

Developing the Voice

The period between ages four and six is a time of exciting maturation, and also a time of waiting patiently for great things to happen. On the one hand, your child's ability to move, sing, keep to the rhythm, and express himself are growing by leaps and bounds. On the other hand, the neural development that enables him to look at written letters and sound them out, to read notes while playing an instrument, and even to coordinate motor programs and visual skills sufficiently to catch or kick a ball has quite possibly not yet kicked in. The point at which this development takes place is not a factor of your child's intelligence. It depends on physiological development, the maturation of a small part of the brain's parietal lobe at the junction where all the senses come together. In most children, this shift in thinking takes place between ages six and eight. Until then, their needs are much better served by focusing on continued rhythmic development, auditory stimulation, and early social skills, rather than by forcing them into chairs and teaching them to read.

This makes sense, when you consider the obvious fact that throughout most of humankind's development, mathematical and phonics notation as well as school desks did not exist. Your child's brain is designed to help him adapt to the environment in which he finds himself, but it isn't specifically designed to keep his eyes on a page and his body in a chair at age five or six.

Originally, kindergartens were created to accommodate just this transition period between the baby years and childhood and to begin to introduce the child to his culture and to the larger world in *preparation* for the academic learning that would come later. Over the decades, this sensible goal has too often been forgotten in the anxiety over reading scores and the perceived need to cram as much information as possible as quickly as possible into children's heads so they will be fully prepared for adulthood. Children as young as five years are too often expected to "sit down, be quiet, and yearn," instead of to actively learn. Reading exercises and drills have been instigated at an age when, for many children, they are ineffective. Children need to move their bodies—to work through new knowledge in whole-body ways, to express themselves physically, even to listen with all their body parts engaged. Too many educators have forgotten how much more easily a child learns to read when he sings and moves to the words he encounters, and how efficiently he learns to write when his body has already internalized the letters' shapes and sounds. Of course, it's important to remember that music must be used wisely to encourage good reading. While it provides the perfect jump start for beginning readers, quiet, undisturbed time with books is also essential for developing reading skills.

Fortunately, recent research reinforces the old assumption that for children this young, learning can best be accomplished if the child's body is involved in the process. One such study reported that three-year-olds who simply attended twice-weekly singing lessons for three years performed better than children who didn't in the areas of abstract conceptual thinking, play improvisation, originality, verbal abilities, and physical coordination. As Dr. Frank Wood, chief neuropsychologist of the Bowman Gray School of Medicine, says, "Music is a primary language of the brain, so to invoke its forms in early childhood is to invoke a muse who already feels comfortably at home in the child's brain and mind." Kindergarten and elementary school administrators are starting to act more frequently on the obvious fact that learning number patterns while marching, fusing numbers to rhythm, is a *pleasurable* experience and hence more productive than sitting at a desk adding two and two. Likewise, at ages four through six, the concepts underlying geometry may be best remembered through simple dance patterns; phonics best comprehended through singing and otherwise making music; and abstract ideas such as gravity, buoyancy, and friction explored through direct manipulation of objects. Meanwhile, the stimulating rhythms of music and movement teach children the *joy* of learning, surely the most important lesson any child can learn at any age.

Many five-and six-year-olds, particularly those who are verbally adept, appear to have already mastered such concepts as phonics, simple mathematics, and even musical notation. In most cases, though, these apparent skills rely on rote memorization rather than true understanding. Until your child experiences his next great period of neurological growth, between ages six and eight, it may not be very productive nor very inspiring to introduce him to ideas in a purely mental fashion through flash cards, memorization, verbal drilling, or any other passive, non-movement-oriented instruction. Your effort as a parent is better spent singing, dancing, and playing with him and, equally important, talking with his new teachers about whether music and movement are integrated into their students' daily routine.

HEARING ANOTHER'S SONG

The first skill we all must master when interacting successfully with others is, of course, listening well. In Latin, *listening* is expressed as *ob audire,* to reach out, to make a conscious effort to connect, to bond, to hold. Through active listening, your child is able to integrate external stimulation into his inner world, that is, he is able to learn. Listening well also tells others that we care about them. It is the first step in initiating and cementing friendships.

A 1975 study by Elyse K. Werner revealed that the typical adult professional spends nearly 55 percent of his or her time listening. Children under eight spend even more time in this pursuit. You have already helped your young child improve his ability to perceive, comprehend, and integrate what he hears through listening games and exercises. He can now learn to refine his listening skills even further, to begin to pay conscious attention to the emotional and informational content of his own and others' speech.

"I don't get it," one mother, a musician, told me after receiving a bad report from her son's first-grade teacher. "Mack seems to pay attention to the sounds all around him. When we go on walks, he's always pointing out the birds' songs or other sounds I haven't even noticed. And I know he loves music. He makes me play my cello for him all the time. But when the teacher or some other adult talks to him, he hardly seems to hear her. His teacher says she has to go right over to him and touch his shoulder before she's sure he hears what she's teaching. It's like there's a wall between him and everything she says."

Even when a child is eager to listen, two barriers may stand in his way in a classroom setting. First, he may suffer from some form of hearing loss. This is not as rare an occurrence as many parents think. In fact, an estimated sixty million Americans suffer from this disability. Due to the noise in our environment, the average twenty-five-year-old in North America hears less well than the average sixty-year-old in traditional African society. Second, the ambient noise in the classroom may be loud enough to interfere with your child's hearing. If your child seems to be having trouble hearing and the children in his class are confined to desks, have him moved to the front of the room near the teacher. Placing him so that his right ear is angled toward her might help, since that is the ear that most efficiently carries sounds to the language centers of the brain. Certainly, it's important to have his hearing checked as well. Even at this age, much preventative work can be done.

If his hearing is fine and his placement in the classroom is ideal, though, such a child's challenge might be one of attention. He may be *hearing* information without being able to *focus* on it, and thus not retaining what he hears. Listening entails the ability to reach out; we listen not only with our ears but with our eyes, our emotions, and our intuition as well. One way to begin to tell whether your child is listening well is to observe how he speaks. When he gives you information, is it clear and correct? What is the emotion within his voice? Is it stressed, natural, or relaxed? Is it rapid, flowing, or disconnected? What is his posture like? Is it stiff, or sloppy and bored? A good listener is almost always a good speaker, so if you feel your child is not communicating very well, you may want to take a closer look at his listening skills.

Perhaps your child has no trouble absorbing factual information he receives verbally but finds it difficult to discern others' emotional state from their speech. He may, for example, fail to note when another child has lost interest in a conversation and wants to end it, or when a child is sad and wants sympathy, or is excited and looking for someone with whom to share that emotion, or when a teacher is becoming annoyed, etc. Of course, it's not fair to expect perfect discernment from four- to six-year-olds especially when we don't even get it from our adult partners. Still, working a bit now on sensitizing him to emotional content can help him toward greater academic and social success in the coming years. Mozart's delicious operatic arias can come in handy here. While driving in your car or relaxing at home, play a recording of "La ci darem" from *Don Giovanni.* To warm your child up, ask him to listen to the oboes and bassoon. Then help him come up with words that describe its mood or tone. If he's interested, the two of you might even create some dialogue to go with the music. Similar games can be played with Mozart's exciting March No. 1 (K. 335).

Another fun way to explore the emotional content of sound is to recite the following nonsense lines to your child:

Osso matta muro zuzu
Frapu fifi neno papu.

Each time you recite the lines, express them in a different way: jubilantly, excitedly, mysteriously, or impatiently. After each recitation, ask your child how he thinks you felt when you said the words. Then reverse roles, having your child recite these or his own nonsense lines in different ways, and guessing how he felt when he was saying them. Once the two of you have tired of the game, take a moment to point out to him how effectively the voice conveys emotion, and how much is communicated through tone rather than words.

Ideally, you will observe a release of tension in your child as he moves through the emotions of the sounds he makes and those he hears. Afterward, he is likely to appear refreshed and ready for further interaction. He will probably also be more responsive as you revert to talking in an ordinary way.

Volume is another aspect of verbal communication that can sabotage social relationships for those children who have not yet learned to truly listen. We have all seen how other children will tend to avoid a boy or girl who always seems to be shouting and ignore the child who speaks so softly no one can hear him. If you suspect that your child falls at one or the other end of the spectrum, casually discuss with him the importance of monitoring one's own volume level in conversation. You will probably have to repeat this discussion a number of times. Later, when your child is talking too loudly or softly, a quick, playful vocal mirror of his sound level may snap him into awareness. Of course, the point is not to ridicule him or make him self-conscious, but to help him feel the effects of his voice within his body. Only then can he correct his error and begin communicating more effectively.

A MUSICAL RECIPE

THE POETRY OF SOUND

Once you sense that your child's ears have been awakened—that he is becoming increasingly aware of the nuance and beauty in the sounds around him—you can help him further appreciate the power of listening with this enjoyable bedtime game. Once he's in his pajamas, has had his bedtime stories, and is snuggled under the covers, ask him to close his eyes and:

"Imagine the sound of . . ." Mix and match the following.

a snowflake	dreaming
a breeze	humming
a candle flame	laughing
a grain of sand	calling home
a rainbow	speaking
a volcano	sneezing

an ice cube
the ocean
a river
a ray of light
a cave
a Popsicle
a drop of rain

whistling
eating doughtnuts
sleeping
weeping
singing
playing a banjo
breathing

Falling asleep to this imaginary music will help your child begin to connect the inner world of his thoughts to the sounds of the outside world. It will demonstrate to him that another's words can powerfully affect his own imagination. Meanwhile, by bringing these charming images to his mind, you are enjoying a moment of closeness with him that he will surely never forget.

TALKING TO THE RHYTHM

"When I taught in the public schools of Chicago, there were two first-grade classrooms next to each other," writes music educator Nick Page. "Fifteen minutes before lunch, the teacher in the first classroom began yelling, 'All right class, put away your pencils. Put away your papers and line up at the door.' Thirty seconds before lunch, the teacher in the second class chanted the words, 'Put away your pencils.' The students echoed the chant: 'Put away your pencils.' The teacher then sang the words 'Put away your papers.' The students echoed back. The teacher finished with, 'Line up at the door,' with the students repeating. By the time the bell went off, her students were ready to go to lunch, and in the next room, the teacher was yelling, 'I told you fifteen minutes ago to line up at the door! Why has this taken so long?' " Clearly, the difference between those two classes was that one teacher had no idea how effectively rhythm and tone can be used to connect to children—and to help them connect with others—and the other one did.

Throughout this book, we have explored the ways in which rhythm and tone inform our physical movements, our emotional inflections, and every utterance we make. Now, as your child enters the larger world of school, you can help him use rhythm and tone to incorporate the natural give-and-take of ordinary conversation into his social interaction; to meld his personal rhythms with those of others; and to cope more successfully with the social complexity of the classroom.

The positive effects of rhythm and tone on young children's school lives have been demonstrated through a number of scientific studies. One such experiment conducted in 1996 revealed that playing folk or pop background music in an ordinary preschool classroom significantly increases social interaction among the children. In a teaching technique called Improvised Musical Play, music is used to draw developmentally delayed children into a play area and engage them in exploration and experimentation, ultimately ritualizing their play format and allowing them to relax and learn. We are all familiar with this kind of social prodding through music. It's why we put on a CD before our own guests arrive. As we move about the room, talking with one person and then another, the gentle or faster rhythms infiltrate our verbal interaction, affecting our communication on a visceral level. It is this sort of musical influence that can be harnessed to help shape the habitual rhythms of communication that our children use. Learning to stop and start on the beat helps integrate a natural responsiveness to all sound, including speech. "Call and response" songs emphasize the pleasure that comes from communicating in turn. Singing together instills a deep sense of harmony and oneness that your child will learn to strive for when interacting with his peers, his teachers, and other adults. And, of course, a healthy diet of new songs helps increase his vocabulary so he can express himself with ease.

Learning to spend time alone recharging is just as important for developing good social skills as is learning to interact well. After a day at school, your child's nerves may be just a bit jangled from the many encounters he's experienced with others. You can help create some healthy downtime for him by consciously setting the mood for his arrival home from school, or yours from work, through music. First, conduct an inventory of the sounds that welcome your child in the evenings. Does a sibling or other family member have the TV or stereo on loud? Are distracting, noisy computer or video games being played? Is the social atmosphere in your home tense and stressful or relaxing and joyful? To help your child regain his equilibrium, orchestrate a restful environment with a quarter hour of Debussy, Chopin, or Schumann's piano music. Such music can provide a wonderful transition between your child's public and private lives.

It is interesting to watch your child move from the musically creative or experimental preschool years to the more socially aware state of the kindergartner. As he moves from age four to five, his musical improvisation will probably begin to decrease just as he begins singing known songs with greater accuracy. By age five or six, he can manage the shape of the melody with the words more efficiently. He has begun to gain control of his voice and use it expressively. Though he's still more likely to sing in tune on his own than in a group, he is well on his way toward social integration through music.

One way to be sure that your child is beginning to internalize the rhythmic instincts that are so important for his social development is to accompany him to a music or dance class. Note what he does when the music starts. How long does it take for him to click in with the rhythm? Does his mouth move listlessly along with the lyrics, or does he actively sing the words? When does he stop singing—before the end of the music, after it ends, or right at the stopping point? Again, it is important to understand that no child this age can be expected to always respond on cue, either musically or in conversation. But now is a good time to begin to understand what his communication strengths and weaknesses are. If he strikes you as still a little off the beat, chant together with him more often than you have been, inviting him to join in, for example, as you tap out a rhythm on your thighs, then on your head, then by stomping your feet, clicking your tongue, and blinking your eyes and finally combining two or more of these actions. Singing favorite songs together, unaccompanied by recordings or musical instruments, should also help him begin to develop the sense of timing that tells us when to start, when to stop, and how to stay together. Keep singing until you're starting and stopping on the beat. Remember, this should be a fun game, not a disciplinary exercise or a drill—pleasurable because your attention is focused exclusively on each other.

Satisfying communication involves not only knowing how to stop and start, but also developing the capacity to take turns and understanding when to speak and when to listen. Parents of young children know that learning to take turns is not an easy process, in conversation or any other activity, and usually isn't completed until first or second grade. However, music can be of great help by allowing children to practice surrendering the spotlight, accepting it, and surrendering it again.

TUNE IN, TUNE UP
SHARING THE MELODY

While driving in the car or just relaxing, initiate some call and response songs that spotlight first one person, then another (i.e., that invoke taking turns). For example, you and your child can sing together:

Blue, blue, blue, blue,
Dad is wearing blue today.
Blue, blue, blue, blue,
Dad is wearing blue.

Look at Dad and we will say
Dad is wearing blue today.
Blue, blue, blue, blue,
Dad is wearing blue.

Help your child to respond with:

Yel-low, yel-low,
Doug is wearing yellow today.
Yel-low, yel-low,
Doug is wearing yellow.

Look at Doug and we will say,
Doug is wearing yellow today.
Yel-low, yel-low,
Doug is wearing yellow.

By doing this, your child experiences that mild flush of excitement as attention turns to him, and he learns to take pleasure in yielding the spotlight to someone he loves.

MOVING IN SPACE

"It's snowy January, and playground activity is at a minimum," Carol Kranowitz writes. "In the music and movement room, I have set up an indoor obstacle course so that our preschoolers can get the touching and moving experiences each little body needs every day. Balance beams, tunnels, ramps, and bridges are among the challenging obstacles. In the background, recordings of Mozart, Chopin, and Bach play to encourage the children's learning.

"It's our second week of miserable weather, and every class has been in the music room at least once. This morning, I'm running a little late. Everything is in place, but the music is not yet playing. But the Butterfly Class is ready, and the children arrive for their scheduled half hour. Kimberly, as usual, jumps to be the first in line. I say, 'Hello, everybody! Here's the obstacle course again. Kimberly, I see that you are ready, so you may begin.' Poised at the tip of a balance beam, Kimberly pauses. 'I can't,' she says. 'You can't?' I ask. 'But you're good at this.' Kimberly says, 'You know I can't begin until the music starts!' "

Just as rhythm and song help children make the inner-world/outer-world connection necessary for everyday social interaction, so the mind-body connection created by music and movement teaches them how to physically express their emotions and ideas. This is important not only for effective communication but also to stimulate the kind of motor development that leads to purposeful, planned movement and organized thought. As your child dances and moves, he expands his awareness into the space around him, opening his eyes to his place in the universe and learning which physical placements and postures keep others feeling relaxed and comfortable. One way you can encourage this development is by playing echo games in which your child mimics your words and movements. While patting your head, hopping on one foot, or performing some other action, for example, chant:

This is what I can do.
Everybody do it, too.
This is what I can do.
Now I'll pass it on to you.

Your child can come up with a different action and pass the ball back to you. It is amazing how much young children love this game, which gives them a sense of control and originality as they learn to sing the correct words at the correct times. It also gives them practice melding physical movement with abstract thought—an important step in learning to experience the visceral felt sensations of new ideas, logical patterns, and profound truths.

We have all experienced physical moments of recognition from a chill at the approach of a threatening stranger to an inner sensation of excitement in the presence of someone especially interesting. Similar physical jolts tell us when we have come up with a great new theory or encountered a valuable point of view. This kind of intellectual attunement with our bodies leads to greater intuitive powers and self-confidence, as our bodies help us discern a truly beautiful intellectual pattern from an imperfect one.

Of course, no one needs to teach a child that thoughts can be felt in the body (he no doubt feels bubbly with excitement the night before each birthday) or that the body can express emotion (just look at his expression and his posture when you tell him he can't have another ice cream cone). Your role as a parent is to teach your child to become *conscious* of the link between his mental processes and his physical expression, so that he can learn to use words to express his feelings and sensations, and his body to help express his ideas. Whether it's free-form movement to a favorite recording at home, ballet lessons at the local Y, a movement program at his school, or the largely child-determined dance of early childhood music education programs, dance shows him what his body is capable of, encourages his language development as he and his classmates create movements together and then discuss them, and stimulates the mind-body connections that will bring his innermost thoughts and feelings out into the light of day.

Mind-body games such as this one, adapted from my book *Introduction to the Musical Brain,* can also help deepen the link between thought and movement. Create four stacks of index cards: one in which each card has written on it the name of a body part, one containing action words (*walk, gallop, slide*), one with words describing the manner in which the action should be conducted (*slowly, heavily, freely*), and one stack containing the names of destinations (*the sky, the hall, the piano*). Then pick one card from each stack and call out a command (such as,

"With your *head, walk* very *slowly* toward *the sky.*"). Your child will delight in the challenge of expressing these ideas and will have even more fun if the two of you take turns.

MUSICAL COMMUNION

"Bobby was five years old, large for his age, and a terror in his Sunday school class," writes Kindermusik teacher Eileen Auxier of Memphis, Tennessee. "If the Sunday school teacher and other children didn't give him his way, he made them wish they had. Sports were his only interest. The director of music in Bobby's church urged the parents of the church to enroll their young children in her Kindermusik classes as well as the children's choir. Bobby's parents enrolled him in both. To say that Bobby was easy to teach would be untrue, but with patience and persistence, the teacher managed to keep Bobby involved in two years of Kindermusik classes, the full amount offered for his age.

"A year or so later, Bobby's parents divorced, and Bobby moved with his mother to a neighboring town. Many years passed. Then one Sunday morning, the teacher noticed a vaguely familiar-looking teenage boy sitting in the congregation. It was Bobby! After the worship service, Bobby greeted his former music teacher with a big hug and said proudly, 'I've still got my glockenspiel.' Bobby's mother added, 'Several times when I have been cleaning Bobby's room, I have asked if I could give away his Kindermusik bag and materials, but the answer is always, 'No, I want to keep them.' Bobby recently graduated from high school, where he was a member of the football team and the heartthrob of all the girls. He didn't become a musician, but his success in school and his good relationship with his peers speak bushels. Music certainly made a difference in his life, as evidenced by his greeting years later to his first music teacher: 'I've still got my glockenspiel!' "

There is something about music that invites people to drop their personal preoccupations and celebrate together. Children are no exception. Most four- to six-year-olds thrill to the excitement of watching a parade or playing in an improvised family band. Group music making is also the best way to hand down our musical culture, our society's collected wisdom, from one generation to another. Originally, one of the primary goals of the kindergarten programs developed in Europe and the United States was to teach children the simple, playful, yet truly wonderful songs of childhood as well as high-quality folk and classical music, and the joys of creating music together.

The neurological, socioemotional, and perceptual-motor benefits of group music making have long motivated school administrators in Denmark, Italy, Germany, the Philippines, and many other countries to routinely include music instruction in their schools. As these benefits are increasingly supported by scientific studies, preschools and kindergartens in the United States have begun reinstating their own music programs as well. Ironically, even as music is being added to preschools across the country, music and arts programs are being cut at our elementary and high schools, as they are deemed and dismissed as a cultural frill. We can only hope that the now solidly documented success of early-childhood music programs will lead to a renewed faith in the power of music to enhance the lives and minds of older children as well.

Generally, the role of the kindergarten music teacher is to awaken children's ears with song; to kindle their interest in music; to begin to mold their musical taste; and to introduce them to their musical heritage. Ideally, four- to six-year-old children should have the opportunity to enjoy singing together, to learn to sing clearly, and to hear a wide variety of children's songs. In this way, the joy of making music together will lead to improved listening abilities, rhythmic sense, and harmonious movement, as well as an enhanced sense of community.

One of the foremost promoters of music education for kindergartners has been Zoltán Kodály (pronounced Koh-DYE), a Hungarian composer and folk music collector who developed a song- and movement-based curriculum for children this age. If you see your child sitting with his class making hand movements to sung scales and recorded music, his teacher is probably using the Kodály method. Kodály believed in the value of folk and traditional songs in connecting children to their community, perpetuating the community's values, and opening children's hearts to the people's most essential ideas and moral strivings. Claiming that a single experience can often open a young soul to music for a whole lifetime, he used the simple pentatonic scale of childhood nursery songs to introduce children to a world literature of music, from Gregorian chants to Debussy. Having spent a great deal of time observing the ways children learn language, he designed a mother-tongue approach to music education that parallels this experience, beginning with informal conversation and working up to symbols and writing—all within the children's own cultural context.

Kodály considered it a parent's moral duty to provide frequent musical experiences to their kindergarten-aged children, especially those that involve folk or children's music. As folk music specialist and academic John Feierabend writes about the Kodály method, "The songs and rhymes of our grandparents have demonstrated community endorsement. They are excellent examples of wonder, are an excellent marriage of words and music, and are still delicious after many singings." Kodály, he adds, understood the importance of music that was born of inspiration rather than financial gain.

Whether your child's school implements the Kodály method or one of the many other excellent music education methods available to teachers, he will always enjoy the experience of joining with others in song. Community music making is a valuable part of growing up grounded and healthy. Whether or not there is a music program in place at your child's school, consider supplementing your child's school experience with an outside early childhood music program in which you participate as well.

SPOTLIGHT ON THE SPECIALIST

SUZANNE TIPS OF KINDERMUSIK

As an early-childhood music instructor of twenty years, Suzanne Tips has seen an endless succession of young children blossom creatively in her Tulsa, Oklahoma, Kindermusik classes. "A wide variety of musical, visual, and movement activities provide each child with the opportunities to strengthen their emotional and mental world," she points out. "With my own style of imagery and movement with music, I can sense if a child is hearing, listening, and developing a good sense of space." One boy, David, was unable to easily participate in her class. As they explored high/low, loud/soft, and fast/slow concepts, Tips began to realize that he was not perceiving the sounds as other children did. His hearing was tested, and when it was discovered that he had a hearing loss in his left ear, Tips was able to show him ways to compensate for his hearing loss and participate fully in the normal class routine. David even created his own instruments, which gave him a means for personal expression.

Music composition is an important facet of the Kindermusik classes. Through Graphic Notation (dots, dashes, and waves drawn in a variety of colors), Tips helps students find ways to represent sound on paper without using the traditional staff and notes. "Drawing sound" encourages children to think about music in visual, aural, and physical ways, she explains, thus integrating it further into their minds, bodies, and modes of self-expression.

FIVE EASY PIECES

Of course, kindergarten and first grade are not all about social encounters. There are concepts and concrete tasks to be learned, too. Already, your child's social experiences have helped to increase his vocabulary. By age six, the typical child has a vocabulary of about fourteen thousand words. The songs of childhood have made him more aware of the passing of the seasons, the days of the week and the months of the year, and the significance and beauty of certain holidays. Making music in groups has introduced him to the wisdom of previous generations. His mind is now wide open to new ideas and capable of picking up and retaining a dizzying number of facts, words, and statistics. With his firm grounding in music, he is ready to be introduced to such new challenges as reading, writing, and mathematics.

As noted before, a four- to six-year-old's learning is most likely to take place in the context of the body. For a child this age, the physical expression *is* the idea. Numbers *are* rhythm and pattern, letters *are* sound. This relationship between body consciousness and intellectual activity in children this age so interested Dr. Marjorie Corso, a physical education specialist in Colorado, that she conducted a three-year study of three- to eight-year-old children. She noted that when she asked the children to touch their shoulders, some touched only one shoulder. Similarly, some children reached with only one hand when asked to jump and touch the ceiling. Later, when reviewing the children's papers, Corso discovered that the children who failed to use a particular quadrant of their bodies avoided the corresponding quadrant of paper when writing and coloring. She also found that the children who had trouble establishing a personal space, lining up too closely to people in front or back of them, usually crowded their letters on the page in the same way. Finally, the children who could not successfully cross the vertical midline (that is, they could not coordinate their movement once one hand had crossed over to the opposite side of the body) tended to stop reading at the middle of a page.

A MUSICAL RECIPE

CROSS CRAWLING

Until about second or third grade, most children are unable to successfully cross the vertical midline, that is, they are unable to perform a challenging task if it involves using the hands or feet in a "crossed-over" position. This is why younger children can more easily learn to play the recorder, but not the fife or horizontal types of flute. An inability to cross the midline that persists beyond this age can eventually lead to certain kinds of learning disabilities, however. According to neurophysiologist Carla Hannaford, author of *Smart Moves: Why Learning Is Not All in Your Head,* the continued delay is frequently the result of the child's having missed some important steps in his physical development—most frequently the crawling stage. To remedy this difficulty, she recommends "cross crawling" to rhythmic music. Cross crawling involves prancing in place to the beat and raising each knee high enough to touch it with the opposite hand or elbow. By cross crawling three times daily for a period of six weeks, using both fast and slow movements to music, your child's body can actually carve the new neuronal pathways in the brain that he needs for optimal coordination and learning.

Clearly, new knowledge must be integrated into the muscles of young children as well as into their minds. It must interact with their senses if it is to be retained in their memory. Teaching musically, through song, melody, tone, and rhythm, is an ideal way to stimulate your child's mind through his body. This fact was demonstrated in a study in which researchers attempted to teach the names of body parts to four groups of four- and five-year-olds. The first group was taught in the standard verbal manner, without music; the second group was taught through verbal instruction along with acting out movements; the names of the body parts were presented in song to the third group; and in the fourth, the new terms were acted out in the form of a dance to music. After twenty days, the three latter groups exhibited higher test scores than the first group. The group exposed to the music-and-dance method showed the greatest improvement both in learning about body parts and on separate tests of creativity.

Music can be especially useful in helping your kindergartner or first grader learn to read. Reading readiness involves a number of distinct skills, including the ability to visually recognize words on a page; the ability to understand that visual parts of words correspond with their spoken sounds; and the ability to recognize words immediately, without having to sound them out. Music is most effective in helping children with the latter two skills. For example, children who do well at telling whether pairs of musical notes or chords sound the same or different also tend to be successful at sounding out written nonsense syllables written on cards—that is, relating the written symbols with the sounds they represent. The more your child has listened to music in his earlier years, the better he is likely to be at discerning the differences among sounds.

Music can also affect reading accuracy, the ability to correctly identify words instantly without having to sound them out. In a 1994 study, twenty-seven kindergartners took part in a group reading program. The group reading was conducted in three different ways: one class sang the text in their books to a musical accompaniment. The second group read the text and also sang it. The third group only read the text. Afterward, the children were tested on how accurately they could read the text without substituting or omitting words. The results showed that the first two classes (the singers) read more accurately than the third. Apparently, music helped the students read because it provided structure and helped them remember what they had read.

Your child's lifelong experience listening to, moving to, and creating his own varied music has equipped him not only to memorize words on a page but also to read creatively—sounding out words, adding to his vocabulary, and increasing his knowledge in a competent, independent way. By continuing to share great music with him, you will continue to stimulate his hearing, speaking, thinking, and reading abilities. This is also a time when the musical improvisation skills that you and your child have honed over the years will come in handy. As he begins to encounter the sounds that letters make on the page, help him play with these concepts by inventing a phonetic "Alphabet Song." You might start him off with something like this:

Ah, ah, animal. Buh, buh, buck.
Cuh, cuh, camel, Duh, duh, duck.

As he joins in, coming up with the sounds that each letter makes, help him provide a word that starts with that sound. The delight he takes in this music making will speed him along the process of learning to read.

At about the same time that basic reading concepts are introduced to your child, or at a number of schools, even before reading is taught, he will begin to practice writing letters on a page. His years of experience in conscious movement and dance will help him begin to master this skill. As with many physical challenges, it's best to start big in helping your child feel the letters of the alphabet, and gradually move toward smaller, more refined movements. One way to do this is to draw the letter on his back with your finger and have him guess what it is. Another is to make a game of forming the letters of the alphabet with your bodies. You can work together to form one letter, take turns forming the same letter, alternate letters, or have one of you form a letter and get the other one to guess which letter it is. You can even get the whole family together to physically spell a word! Activities such as this increase your child's awareness of each letter's shape, an understanding he can then move into the finer muscles of his hand.

During my years spent teaching in Japan, I witnessed another version of this moving-from-large-to-small method of teaching young children to write. At one school, children were introduced to the writing of Japanese script while the recorded music of Mozart and Debussy played in the background. The teacher provided the young students with large sheets of paper fastened to the walls, along with large brushes and paint. They were asked to re-create specific characters on a giant scale on the sheets of paper, using their entire bodies to create the large sweeps of the single character. The teachers did not judge them on the quality of the work, but simply required the children, day after day for about a week, to continue acting out the strokes of the character with their bodies. The next week, somewhat smaller sheets of paper were provided, with chalk instead of brushes, and the children re-created the character with whole-arm movements rather than with their entire bodies. In the third week, the children sat at their desks and drew the characters with large, soft-lead pencils on unlined paper, limiting their visible movements to their hands and wrists. By the fourth week, when they were given a regular pencil and paper, they were accurate and relaxed in their fine-motor movement due to the creativity and flow internalized by their gross-motor experience. I was impressed by how accurately, and how quickly, the children learned to write with this method.

Your child can progress in a similar manner as you move from creating whole-body letters of the alphabet to a painting easel. Show him how to paint a letter that fills the entire sheet of paper. Hum or sing "Uuuup and doooown and aroooound" as you gently move his hand in the proper directions. The following week, have him repeat his actions on a chalkboard, if he has one. The week after, have him sit at a table with a large sheet of paper and a thick pencil. Be sure to play light, well-organized classical music softly in the background as he works. This may facilitate his spatial awareness and help him focus. With practice, your child will begin to draw reasonable facsimiles of letters of the alphabet and have plenty of time to think about what a written letter actually means while he works.

Music pervades the entire being of a four- to six-year-old, extending its nurturing arms not only into the areas of emotion, social interaction, cultural grounding, and reading readiness, but to the earliest stages of mathematical learning as well. Playing an instrument, for example, has been shown to increase sequencing skills (and therefore mathematical abilities) and adeptness in logic, according to Pierre Sollier, director of the Mozart Center in Lafayette, California. The stimulating effects of an instrument's high frequencies can also boost focus, attention, and memorization and provide a greater energy to learn. One group of nearly a hundred remedial first graders, for instance, was given seven months of music instruction and visual-arts training. By the end of the seven months, they had caught up with an average-performing group of children not trained in the arts, and actually surpassed them in math. The researchers weren't sure whether the pleasure of arts classes motivates kids to do better, or whether music exercises cause a kind of mental stretching. Whatever the reason, the process worked.

As your child moves from physical awareness toward concrete thought, it is important to keep in mind that music continues to positively affect the ways in which his brain operates. A group of Russian investigators recently demonstrated that this benefit occurs even in preschool children. After being exposed to classical music for one hour each day over a six-month period, a group of four-year-olds showed an increase in alpha rhythm frequency associated with a relaxed, daydreaming, creative state and greater coherence, or cooperation, among different regions of the brain. It is interesting to note that the children were not asked to listen to the music. It simply played in the background while they went about their activities. And yet it had this measurable effect. Billie Thompson, director of the Sound Listening Center, has long made use of this phenomenon in helping children at her Tomatis-based centers in Phoenix, Arizona, and Los Angeles. "Mozart is a key ingredient in teaching people how to listen and learn effectively and efficiently," she points out. By moving, drawing, and playing as they listen, children enhance their listening experience.

In short, the way to a child's mind is still through the senses. By providing your kindergartner with rich, harmonious, *musical* learning experiences, you are stimulating and enhancing his capacity to think in a clear and increasingly complex manner. By introducing his body and heart into the wonderful world of mind through music, you are showing him what a pleasure learning can be.

FOR YOUR CHILD'S HEALTH
HOSPITAL HARMONY

The preschool, kindertgarten, and early elementary years, when children are freer to run, play, experiment, and socialize, are sadly a time when accidents and illness are especially common. If your child is faced with a stay in the hospital, you can use music to ease his stress and even lessen the pain of surgery and other invasive procedures.

The loss of control and overstimulation of sounds, medical equipment, and unfamiliar people make the pediatric intensive care environment especially frightening for many young children. Studies have shown that playing recordings of music your child enjoys can create positive *physiological* effects, including decreased heart rate, less rapid breathing, and lower levels of the stress hormone, cortisol. Music can also distract your child from his stressful environment and create a private room of sound that can enclose him (and you) in its healing warmth. By aligning himself physically and emotionally with the serene rhythms of Mozart or other classical music, your child may also be able to speed up his recovery. In general, slow, soothing lullabies, recordings of bedtime stories, and light and beautiful songs that don't overstimulate are best for calming a child who is anxious about his condition or an upcoming treatment. At times, though, your child's favorite upbeat songs may create a perfect distraction from pain. Offer him a variety of selections from within these categories, and let him decide which songs work best.

MUSIC: THE REMEDY FOR CHILDHOOD STRESS

"The first thing children do when they enter the preschool building is to hang up their coats," writes music educator Carol Kranowitz.

"Since school started last month, Wylie has not been doing this. Moving dreamily down the corridor, he frequently loses his way and takes detours, or, if he does make it all the way to his hook, he just stands there. Wylie's classroom teachers hover nearby, giving gentle encouragement: 'Wylie, we're indoors now, so you can take off your coat. . . . That's it. . . . Now you can put it on your hook. Like this, Wylie. Wylie, are you listening?' Wylie is an out-of-sync child, and all of us on the staff are concerned about him. One teacher remarks, 'If only we could figure out how to give him a jump start!'

"Today, I'm in the corridor, rounding up Wylie's classmates to come to the music room. The other children are lining up, but Wylie, still in his coat, is at a standstill. I ask, 'Are you ready for music, Wylie?' He looks at me and nods. Otherwise, he doesn't move. So I try a different tactic: I crouch in front of him, look squarely into his eyes, and sing to him. To the tune of the A-B-C song, I repeat the question.

"Are you—ready for—mu-sic—Wylie?

"Suddenly, Wylie comes to attention. His eyes light up. He laughs. He whips off his coat, hangs it neatly on his hook, and leaps into line. WHAT IS THIS? Is singing the way to arouse Wylie? Is he a child whose brain is wired to respond to language that is sung rather than spoken? Astonished, I start the song again, piggybacking it to 'Twinkle Twinkle' and making up rhymes that are inexpert, but good enough:

"Are you—ready for—mu-sic—today?
Let's get—mo-ving—here's the—way.
Now we're—walking—down the—hall,
Here we—go—one and—all,
Are you—ready for—mu-sic—now?
Let's go—in and—take a—bow.

"Wylie participates in the music class as never before. Now, whenever his teachers want to elicit a reaction, they look him in the eye and sing their question or instruction. And what do they get? An attentive, related little boy who is 'in sync' when he not only hears, but also faces the music."

For a wide variety of reasons, many children enter our schools today with an inability to get in sync—to meld their personal rhythms with those of others. This inability to be part of the crowd can create stress in a young child that only increases over time if his rhythmic challenge is not met. Sometimes, the problem lies not with the child but with the environment. The school drills, announcements, traffic noise, or other out-of-sync children create a constant distraction.

These days, few child development researchers seriously doubt music's power to fine-tune the brain. However, not all are as aware of the ways music has been proven to reliably alter young children's moods in the classroom. Helping them overcome anxiety and stress, calming them so that they can focus on a task, and enlivening their bodies sufficiently to maintain alertness, music works its magic whether or not the children consciously pay attention to it. Given evidence now available, it is a mystery why music is not used more frequently in this country's elementary schools.

Music education for children is not a frill. Good music instruction touches the child's mind where it lives—in the body. It is essential for the rhythm, the balance, the emotions, the social awareness, and the increasingly sophisticated thinking of the kindergarten and elementary school child. As a parent, you have done your best to provide the richest, most appropriate environmental stimulation possible for your child. Now it's your school's turn. When you are visiting the school your child will attend, don't neglect to ask whether a music instruction program is in place. If you notice that your school administrators are just beginning to include the arts in the curriculum, share with them the wonderful studies demonstrating how music helps develop verbal, emotional, physical, and academic skills. Keep in mind that many of the techniques you have learned in this book are ideal for home schooling as well, enabling you to instill greater variety in the daily routine and therefore work for longer periods of time. If there is no program, you and other parents might take a cue from the students at Bohemia Elementary School in Cottage Grove, Oregon. After that school's music program was eliminated due to funding cuts, students held a spellathon to raise the $22,000 needed to pay for a part-time music teacher. Now a teacher visits each classroom in the school once a week and directs an after-school jazz choir, all due to the children's efforts to keep music in their schools.

If music, including the sounds and rhythms of children's play, has been incorporated into the everyday experience of your child's school, chances are excellent that your child's needs will be appreciated and understood there. But don't stop with just your own child. Look around your community. Are children of all neighborhoods and socioeconomic strata offered music in your city's schools? Be sure to consider this question when voting in school board elections and other venues in which music-program funding may be at issue. Access to quality music, whether popular or classical, should be the birthright of every child.

A MOZART MUSICAL MENU

- The Rondo from *Eine Kleine Nachtmusik* (K. 525). This rondo is one of the most charming melodies ever written and easily brings a child's mind to attention. Use this music to alert your child's mind that desk time is approaching, and that it can be joyous and happy.
- "Vol che sapete" from *The Marriage of Figaro*. This haunting and beautiful song can be translated as, "You have the answer, you have the key." Its melody allows your child to fantasize about beauty and joy.
- The Andante from the Cassation (K. 63). *Andante* means "walking slowly." An andante gives us a sense of slow movement. Suggest that your child close his eyes as he listens, and imagine what it was like when Mozart, a young child of eight or nine, heard this melody in his head and wrote it down. Perhaps he was traveling in a carriage from one city to another and was inspired by what he saw out the window.
- *Variations on Ah! Vous dirai-je, Maman* (K. 265). Learning to approach desk work with variety and freshness is important. It's best to have a regular routine and enter into it every day with a fresh new mind. Just as this lovely piece stimulated your infant's mind, so it can awaken his brain to new ideas now as he begins school.
- *Die Leyerer* (K. 611). Ask your child to imagine himself walking with young Mozart through one of the beautiful cities of Europe. The two of them come across a hurdy-gurdy man playing his primitive instrument. Encourage your child to listen, dance, and sway to these beautiful and unique sounds.

- *La Bataille* (K. 535). Here is a whole story of a little battle between tin soldiers. Give your child the freedom to make up his own story, play, or pantomime. It's all in the music.
- The Menuetto from the Divertimento in B Flat (K. 287). Minuets are dances that often have a waltzlike beat. Here's your child's opportunity to make up his own ballet, exercise, or stretching movements. Encourage him to use his hands, feet, elbows, knees, and nose while seeing how many ways he can put Mozart into motion.

CHAPTER 8

RHYTHMS OF THOUGHT

Learning, Remembering, and Expressing (Six to Eight Years)

Children can learn almost anything if they are dancing, tasting, touching, hearing, seeing and feeling information.
—JEAN HOUSTON, EDUCATING THE POSSIBLE HUMAN

The years 1763 through 1766 were remarkable ones for young Mozart. During those four years, the sparkling prodigy toured nearly all the courts of Europe, accompanied by his father, Leopold. Leaving home at age six, Wolfgang performed for kings, enchanted princesses, rode in carriages through great cities, and wandered the squares of foreign capitals. He returned home at age nine a full-fledged composer, having brilliantly crossed the bridge from the sensory experience of early childhood to the concrete reasoning of an older child.

Your little one will make a great journey, too, during her elementary school years, leaving behind the simple chants and dances of her early life and entering the vast, challenging world of reading, writing, arithmetic, and (perhaps) music literacy. As her brain makes its third major leap in development, creating brain-to-brain connections that depend more on mental frameworks and less on sensory input, academic work will become an increasingly important part of her life. Of course, her study skills will be rudimentary at first, but music can help create healthy habits that will support her throughout her school years. Rhythm and tone can also help structure her first forays into logical thought, increase her retention of new concepts, and guide her through her early work in reading, writing, and arithmetic. By reinforcing her self-confidence, self-expression, and creativity, it can help form an active, questing, independent mind.

This seems like a tall order for any educational tool—much less one as natural and pleasurable as music. Even Peter Perret, music director of Winston-Salem Triad Symphony Orchestra, was skeptical, if curious, when he heard a 1992 news story on National Public Radio claiming that learning to play an instrument at a very young age increased children's academic performance. Mr. Perret thought it unlikely that the children who most needed stimulation would get the music lessons that might help them, but he wondered whether just listening to music might have a similar effect. To find out, he arranged for a woodwind quintet to play at Bolton Elementary School in Winston-Salem, North Carolina, every school day for three years. The first year, the quintet visited each first-grade classroom for two or three half-hour performance sessions per week. The second year, they played for the first and second grades, and the third year for the first through third grades. In addition, the school principal, Dr. Ann Shortt, played recorded classical music over the school's intercom system so it could be heard in the halls, library, and lunchroom.

The students at Bolton Elementary School were not high achievers initially. At the time that the woodwind quintet arrived, the average composite IQ score among the second graders and fifth graders was 92. These children had had few advantages, and in many cases not much extracurricular stimulation: 70 percent were poor enough to qualify for free or reduced-price lunches. Yet the effects of the quintet's expert music making seemed almost immediate. Three weeks into the first year, first-grade teacher Carol Cross observed, "I can already see a difference in the children's ability to listen."

After three years, the children who had listened to the music were tested. The year before, fewer than 40 percent of the school's third graders (none of whom had been exposed to the quintet's music) had tested at or above their grade level. This year's group of third graders, who had had three years of exposure to the quintet, fared much better.

Eighty-five percent of this group tested above grade level for reading; 89 percent for math. Dr. Shortt was convinced that the musicians had a direct impact on these students' superior achievement scores.

It is a shame the quintet could not stay longer at the school. While so much of the school experience consists of rote transmission of preexisting knowledge, music stimulates the kind of creative thinking and active listening that leads to true learning. Music's beauty and mystery bring a sense of wonder to the learning process, as well. Sitting at a desk or cross-legged on the floor, your child can dive into the serene world of a baroque concerto, and then awaken with a fresh enthusiasm for reading, mathematics, science, and art.

As your young student begins her climb up the academic ladder, rhythm and tone can act as friend and companion as well as her guide. When she is tired or discouraged, her favorite melodies will be there to console her. When she is confused, a lifetime of rhythm will help order her mind. Expose your child now to the multifaceted potential of high-quality music, and it will serve her through all the years to come.

THE MUSICAL BRAIN

Sometime between ages six and eight (usually later for boys than for girls), a period of dramatic growth will occur in your child's brain. The development of a more elaborate set of spinal pathways will lead to the establishment of new links among the brain's visual, speech, and motor regions. The growth is so great that midway through the process, at about age seven, the skull actually gives way and grows in size. The new connections will allow your child to link visual stimuli (what she sees) directly to auditory stimuli (what she hears) without having to process the information through her body first. This greatly increases her ability to think abstractly and to reason logically, and boosts the development of her own inner voice or thought process. It also enables her to begin to understand codes and symbols, like the ones used in math, phonics, and music notation.

The more exposure to music your child has already had, the deeper and more successful this neural integration is likely to become. If rhythm has already helped structure her thoughts, logical reasoning will naturally follow. If melody has attuned her ears to nuance and emotion, intellectual discernment will improve. To continue supporting her growth during the early years of elementary school, it's important to know your child's musical requirements at each stage of development. For first graders, the key is rhythm: moving from simple starting and stopping to an ability to match a steady beat, thereby coordinating the clapping of hands with the mental apprehension of a melody. This mind-body experience of rhythm will eventually lead to more complex skills—listening, processing visual information, coordinating movement, and knowing where she is in time and space.

In second grade, you can encourage your child to begin breaking down the rhythms she has grown to love: to experiment with new patterns, to learn new games and dances, and to explore the auditory spectrum through singing or playing an instrument. These kinds of improvisation will give her more of a sense of herself as an individual as they sharpen her listening skills.

Third grade is a time for your child to get to know her body in a different way, through folk dances, singing, and perhaps participation in a small instrumental group. Songs about transformation (caterpillars into butterflies), fantastic journeys (going to the moon), and heroic rites of passage will captivate and instruct her. Jumping rope and singing epic songs with a hundred verses will instill a wonderful feeling of being part of a whole and pave the way for the self-integration process that typically begins the following year.

Each of these new discoveries will supplement your child's exploration of the world through books, numbers, and an amazing variety of new concepts. By leading her steadily through the steps of music and thought, you can help her make the most of the mind with which she was born.

TUNE IN, TUNE UP

RHYTHMIC CHANTS

As your child grows, it's as important to continue singing and chanting with her as it was when she was younger. The emphasis at around six years, though, should be on rhythm. Organic chants, inspired by the concrete elements around you at the time and improvised on the spot, are still the best. Invent some rhythmic chants with your first grader while waiting in the line of cars at your favorite fast-food place, at bedtime, or in the morning over breakfast. Chants involving hiding objects and then producing them are especially popular with children this age. A typical chant might go like this:

I lost my jingle bell,
Jingle bell, jingle bell.
I lost my jingle bell.
Where has it gone?

I found my jingle bell,
Jingle bell, jingle bell.
I found my jingle bell
In a pocket right here.

Finding the rhythm involves learning to command the body. At this age, thought first begins to control function—the first major step in growing up.

SERENADES FOR STUDY

Mozart loved math as a child. He covered the furniture and even the walls of his bedroom with numbers. While on a trip to Italy, he wrote home to his sister, Nannerl, begging her to send him some math tables he had lost. As an adult, Mozart loved to play billiards with Constanze, his wife. Both math and billiards, along with music composition and performance, require spatial reasoning—one of the very skills your child is beginning to develop now. Given Mozart's passions, perhaps it shouldn't be surprising that his music stimulates this very type of intelligence.

Music, not just performing or composing, but simply listening to, singing, or moving to music, has also been shown to increase reading ability, memorization skills, vocabulary recall, and creativity. In one study, first graders who participated in Kodály music appreciation and movement five days per week, forty minutes per day, for seven months, scored significantly higher on a standard reading test than a nonmusical group—eighty-eighth percentile as opposed to seventy-second.

One reason why this is so may be music's physiological effects on the body. Listening to music has been shown to slow the listener's heart rate, activate the brain waves of higher-order thinking, and create a positive, relaxed, receptive state of mind that is ideal for learning. Advertisers have long taken advantage of music's power to open the mind by coupling ads with catchy melodies that lodge in the listener's memory. Parents and teachers can make use of this same process by reciting learning materials to a background of baroque or classical music. The structure and aesthetic attractiveness of the melodies have been shown to aid in retaining the information that accompanies it. The material melds to the music, as it were, just as simpler information, such as your Social Security number, is easier to recall in a fixed rhythm. This form of learning is hardly new. The ancient Greeks were highly familiar with the effectiveness of reciting words to music. Yet, for some reason, throughout the latter half of this century the technique has largely been neglected in our schools.

Renowned music researcher Jay Dowling believes that music's positive effects on various kinds of learning may also have much to do with its combining of two forms of mental processing. There are two kinds of memory, he explains: *declarative* memories, which can be consciously recalled and stated (your wedding day, the *Challenger* disaster, the multiplication tables), and *procedural* memories, which are learned but can't be described in words or even consciously recalled (how to skip, how to balance on your head, how to read a word). Declarative memories are more mind-related, while procedural memories belong to the body—and music's ability to combine mind and body processes into one experience facilitates and enhances the learning process. Procedural learning influences in a very fundamental way how we perceive and understand the world, Dowling says. By integrating mental activities with sensory-motor experiences (moving, singing, or participating rhythmically in the acquisition of new information), children learn on a much more sophisticated and profound level.

Dr. Carla Hannaford explains that "singing stimulates the nerves to the vestibular system, the eyes, ears, and vocalization areas, thus waking up the brain to new learning and optimally taking in sensory information." Dowling adds, "If instead of talking to children, teachers got them singing, it would help a lot and would produce active involvement. Also, it is fun." Even teachers who don't sing can vary the pitch and rhythm of their voice to make it more interesting to themselves and their students.

LEARNING TO THE BEAT

"The name of our president is MISTER Bill Clinton. The name of our vice president is MISTER Al Gore," schoolchildren in one of Houston's interracial neighborhoods steadily chant. The children sit with hands on their desks, feet on the floor, their eyes on their teacher or on the blackboard, as the teacher claps the beat. Through this extremely rhythmic rote learning method, large numbers of inner-city children learn addition, multiplication tables, verbal math problems, the states and their capitals, vowel sounds, and even reading, in areas including Baltimore; Chicago; Broward County, Florida; and parts of Utah. Rote-like teaching methods, developed during the War on Poverty in the late 1960s, largely abandoned and recently revised, have brought these Houston first graders up to a ranking of thirteenth in reading out of 182 schools citywide. Since then, it has been adopted by a number of schools whose students are substantially better-off.

While I am not in any way advocating a strict, stay-in-your-seat, rote-memorization learning process (on the contrary, I support increased physical involvement in elementary school learning), the results gained in Houston and other such areas do speak to the enormous power of rhythm for children of this age. The beat of the movement, the clapping, the heartbeat, the pulse, the predictable structure of a song, are everything to a six-year-old. The experience of beginning at the same time and ending at the same time is a sublime, visceral pleasure. Much more important at this age than singing in tune is the sense of steadiness, a regularity of rhythm that provides them with an envelope of security, of knowing where to go. Whether they're singing on the bus, jumping rope, or marching in a parade, six-year-olds love to have a clear sense of where they are. It's a magical experience and it's a time when a great deal of learning can take place.

The way such learning occurs is through rhythmic patterns that are embodied in your child through her vestibular system, the melding of the body with the brain. This physical encoding bundles information so that it can be more efficiently processed. In general, a person's short-term memory has the ability to hold about seven bits of information. When related groups of information are bundled through rhythm, they may be remembered procedurally as one bit of information, and the volume of material that can be stored increases. Much of your child's early learning of language was dependent on the rhythmic presentation of words. Multiplication tables, the alphabet, spelling, and poetry can all be memorized in this way as well.

Of course, rote memorization is just a small part of the learning process. The depth of a child's enthusiasm has much to do with how long the results of any lesson last. Making this process more creative, changing the pace, cadence, and intonation of the voice, using new words so as to reinvent the material each time, helps children absorb new knowledge on an emotional level. As we have seen in earlier chapters, emotional associations have long been proven to increase and enhance learning and memory. In other words, it's important to use rhythm, but it's just as important to have fun.

A MUSICAL RECIPE

THE UNNERVOUS TIC

The last thing you want, as your child enters the wonderful world of reading, is to see her bogged down by the tedium of endless spelling lists. By keeping her spelling homework fun, you keep her interest in words alive. Here's a rhythmic game with which to practice her next series of words:

Say the first word on her list is *today*. First announce the word. Then tap or clap a steady pulse. Your child will probably enjoy tapping it with you. As you keep the beat, spell the word to various rhythmic patterns, as illustrated below. After each spelling, have your child repeat it with you. If your child comes up with her own rhythms, so much the better.

Speak: T o d a y
Tap: • • • • • • • • • • • •

Speak: T o ______ d a y ____________
Tap: • • • • • • • • • • • •

Speak: T ___ o ___ d a y ____________
Tap: • • • • • • • • • • • •

Speak: T ________ o d ______ a y _____
Tap: • • • • • • • • • • • •

Now adjust the pitch of the spelling so that each time you spell the word it has a different melody.

Finally, combine both methods of spelling *today* and let the letters melodically sound in rhythm. Addition and multiplication tables can be learned in a similar way.

It sometimes helps when reviewing spelling words to say and spell the word first toward your child's right ear, then toward the left, so that you are sure to stimulate both sides of her brain. Just as most people are strongly left- or right-handed, we tend to listen more with one ear than with the other. For listening to be most effective when communicating verbally, the right ear—with its many connections to the language centers in the left hemisphere of the brain—should have the leading role. In his fascinating book *When Listening Comes Alive,* Paul Madaule recommends a way for your child to explore the ways in which she listens with one ear or the other. First, she should sit on a stool in an erect but relaxed position with her left ear facing the sound source, close her eyes, and listen to a recording of a Mozart violin concerto or sonata. As she listens, encourage her to focus with her left ear on the violin sounds, then turn her head and focus on the sounds with her right ear—then back to the left, to the right, and so on. After two or three minutes of this, have her return her listening to her right ear. Then very slowly rotate the stool to the right, exposing her left ear to the sound source while she continues to listen with her right ear. When she feels that her right ear is losing its lead, slowly rotate her back until her right ear becomes dominant again. By repeating this action for about five minutes, she will learn to consciously control her listening and train her right ear to become dominant. Finally, have her face the music source and listen with her right ear. Then, ask her to imagine that her right ear is moving slowly to the top of her head. In this way, she will listen to the music with both ears again, but the right ear will remain dominant.

In first grade, beginning reading skills are seriously introduced. The process of reading entails a complex synthesis of pattern identification and inner speech, that is, the ability to hear words internally. Your child should be well prepared for this activity after all the listening games you have played with her over the years. However, at age six nearly every first grader is still in the process of developing her inner speech abilities and therefore needs to read aloud and hear herself speak in order to understand the words.

Again, rhythm can help your child master this skill. You might read rhythmically to your child, having her rhythmically echo each phrase. Change your pitch with the rhythm, and have her do the same. When she reads aloud, keep a soft tapping rhythm going to guide her voice. Let her have worry beads, clay, or something to keep her hands busy while she reads. A rhythmic musical accompaniment (preferably string instruments rich in harmonic overtones) has also been shown to improve reading performance over time. Try turning down the lights a bit and playing Mozart, Vivaldi, Scarlatti, or Bach softly in the background as your child reads aloud. Read to her to the same accompaniment as she silently reads along. Then read together, melding your voices together with the soothing rhythms of the music.

Finally, consider the fact that, even if your child's classroom seems quiet to you, it may be loud enough to interfere with her reading and spelling abilities, concentration, and behavior. According to sound scientist Carl Crandell, young children need quieter environments than adults to be able to hear as well. In fact, children don't develop an adultlike ability to recognize speech in noise until about thirteen to fifteen years of age. Studies have shown that even in a relatively good classroom listening environment, children are only able to recognize about 70 percent of one-syllable words. In poor classroom environments, which are more common, they recognize less than 30 percent.

Distance from the speaker has a great deal to do with how well your child can hear in the classroom. Dr. Crandell's 1986 study of children aged five to seven with normal hearing revealed that they hear at 89 percent correctness at a six-foot distance from the teacher in a typical classroom, 55 percent at twelve feet, and 36 percent at twenty-four feet. Small classes seem to be the best way to ensure that children can hear well enough to learn effectively. If your child is in a larger class and is easily distracted, speak with the teacher to be sure she's sitting in an optimal listening position.

TOOTHFALL TIME

Perhaps it's the psychological jolt caused by losing one's front teeth that makes second graders more open to the idea of breaking down the rhythms they worked so hard to master a year or two before. At this age, you can introduce 3/4 waltz time in addition to the old 2/4 and 4/4 of nursery songs. Occasionally, when talking or singing with your child, playfully vary the pulse of the music or words, and encourage her to do so. Varying rhythm and pitch may not be new experiences for your child—you may have played in these ways with your eighteen-month-old—but they acquire new power and significance for children of this age.

Melody, whether it involves listening to music, playing an instrument, or singing, also becomes more important now. Whereas rhythm activates patterns of movement in the body, melody is closer to language through its phrasing and contour. Your child can discover a variety of sound combinations by listening to new music or creating interesting chords on the piano, and explore differences in pitch through such songs as the Beatles' "I Want to Hold Your Hand." He can experience the multiple blends that human voices and musical intruments produce by playing a simple duet with you or humming along to Peter, Paul and Mary. This aural stretching can especially improve your child's ability to sound out words, read aloud, and write—all skills that require a clear, stable, and precise perception of the acoustic content of words. Accurate listening also instills grace, consciousness, and intention in your child's movements, even as it enhances her self-confidence and social skills.

Just as with rhythm, your child is ready to *experiment* with melody now, departing from the patterns that have supported and reassured her through her earlier years. Imitating a fire engine siren helps her learn to vary the pitch of her voice. Singing louder, then softer, then louder again, like a radio being turned up and down, allows her to explore the effects of volume changes. Children's songs featuring funny noises or sudden changes in pitch or volume are popular and beneficial at this age.

In general, the second grade is a time of much improvisation. If your child's school has a music program that involves creative invention with musical instruments, movement, and song (such as the Orff Schulwerk approach), your seven-year-old will feel right at home. The Orff approach, designed in the 1920s by German musician and composer Carl Orff, is based on the concept of *elemental music*—that is, the incorporation of music and movement in one expressive form, and the integration of the natural into "the moving, expressive, auditory world." Today, in a typical Orff classroom, your child is likely to recite poetry or create stories as she moves, claps, and taps out rhythms on a drum, wooden xylophone, or metal glockenspiel. Her teacher may draw on fairy tales, myths, poems, or simple folk songs to help her fully experience the qualities of music and dance.

Such programs, which emphasize the physical aspects of creativity over more symbolic aspects, such as reading or writing music, stimulate your young child's imagination, develop her emotional awareness, improve her active-listening skills, and allow her to interact with others in collaborative, noncompetitive ways. If your child's school has not yet put such a program in place, consider talking to the administrators about doing so.

As your child moves from second to third grade, her perceptiveness regarding the world around her—and the stories she hears—will increase. You can encourage this healthy expansion in thinking skills by playing and singing folk ballads and other songs that tell stories. After singing or listening to such a song, ask her how she thinks a character in the song felt, or why he was happy or sad. Ask what she thinks the characters will do next, after the story in the song is over. Encourage her to connect the song to her own life. ("How are you like that little boy?" "Have you ever done anything like that?") As she moves from sounding out words to reading for content, conversations like this will help make her a better reader.

Naturally enough, children also tend to become fascinated by the idea of transformation and physical change as they reach the end of their early-childhood years. The uneasiness that your child's fast growth and other physical changes can cause can be eased by songs addressing these changes. When she loses a tooth, instead of talking or singing about the tooth fairy, help make up a song about how much the tooth helped in chewing food, how much it will be missed, and how strange it feels not to have it anymore. Songs that deal with the way the body stretches and grows are also useful at this age. Music dealing with transformations in nature—the passing of seasons, the phases of the moon—becomes important now because changes in the natural world mirror changes in the human body.

The eighth year is a time when the magic of communal stories, whether musical or literary, truly begins to take hold as well. Group dancing and singing allow your child to experiment with the ways in which her separate self can meld satisfyingly with a group. Singing songs with many verses, jumping rope, and reciting football cheers allow her to meld with rhythm itself. Joining a chorus or a band can encourage her in this group experience, especially if she tends to be shy or solitary. Remember, these early-childhood experiments with separateness and unity are necessary precursors to a more integrated state. As she learns to be alone and to be with others, support her choices by putting on some music and then leaving the room or, if she's in the mood, putting your arm around her and listening together.

Finally, third grade is a year of increased memorization in the classroom. Don't forget rhythm's power to help your child commit information to memory. Linda Kittchner, a music teacher in San Antonio, Texas, writes of her fifteen-year-old daughter's argument with a friend about the number of states in the Union. "One of the girls said there were fifty-one states," Kittchner writes. "The other said, 'No, it's fifty! Didn't you ever learn that song?' and proceeded to sing 'Fifty Nifty United States,' by Ray Charles. After she finished I asked her when she learned that song. This was a high school sophomore, and she had learned it in the fifth grade, and it was still right there."

Spotlight on the Specialist

JUDITH COLE, ORFF SPECIALIST

As a board member of the American Orff Schulwerk Association, Judith Cole has been teaching music for twenty-five years to children of all races and nationalities in Texas and elsewhere. "Orff Schulwerk helps develop expression and experiential foundation for comfort in a new language," she says. "Young students who don't speak English are able to translate their feelings through the simple patterns of rhythmic speech and movement."

A member of the faculty of Texas A&M University in Kingsville, she now trains students for careers in education and music. She was not especially surprised to note that her adult students, most with many years of formal music training, reveled in the improvisatory nature of the Orff technique. "Orff Schulwerk prepares every student with both improvised and structured tools for assisting in language, movement, and creative expression," she explains. "When they go on to start their careers, poetry, rhythms and rhymes allow them to become more creative and spontaneous in their elementary school classrooms, whether they're teaching music itself or the language arts."

Rhythms of the Day

Just as you learned and helped mold your infant's daily rhythms in the six months after she was born, so it is important now to consider her natural rhythms of physical, emotional, and mental energy, and the ways they affect her life. Though some of the daily up-and-down rhythms in her ability to focus and study effectively will change as she grows, she has already established some basic patterns that reflect her physiology, upbringing, and genetic inheritance. The more you can help her get to know these patterns, the better able she will be to accommodate them, and to work around them when necessary.

Certainly by now you know whether your child is a morning glory or a night owl. Practically from birth, she has probably greeted each day either with great energy and enthusiasm or a groaning desire to go back to bed. If she likes to get up early, her entire morning is likely to be a high-energy time, while her afternoons are better for quiet pursuits. (The opposite is true, of course, for night owls.) Both morning and night people, however, have certain things in common: almost everyone experiences a sudden drop in alertness between 3:00 and 5:00 P.M., and the lowest energy level of all between 3:00 and 6:00 A.M.

When your child needs to concentrate during a time when her energy is down, help her explore ways to fight her natural tendency to droop. Lively music is one way to get more in step with the task at hand. Warming up with movement is also a good idea. Don't forget the effectiveness of starting with slow music similar in energy level to your child's, then gradually increasing that musical energy until she's been brought up to speed. It's even better if your child can invent her own methods for riding the ups and downs of her day. In any case, her awareness that these rhythms are normal and universal will help her accept them in herself and manage them in the most efficient way.

Our daily rhythms are not the only patterns we can find in our lives. Each of us operates according to the smaller rhythms of attention span, rhythms that change as we grow. Your six- to-seven-year-old, for example, can probably pay attention to a task for about seven to eight minutes. An eight- to-nine-year-old's focus tends to last about seven to ten minutes. Attention span lengthens by about two or three minutes every two or three years. At this age, your child needs at least two minutes' transition time before beginning a new activity. These patterns are important when helping children begin to create good study habits. During third grade, homework is likely to become noticeably more challenging. By keeping an eye on the clock, you can estimate when her attention is likely to lag. When you see that she is beginning to lose her focus, give her a stimulating "sound break" by playing a recording and encouraging her to move or sing along to reawaken her interest in her work.

FOR YOUR CHILD'S HEALTH

EAR GLASSES

"I wanted to let you know that Mozart got me through school and grad school," writes a student at the University of Massachusetts. "I wrote dozens of papers with Mozart on the headset. I have a severe learning disability that impairs my ability to write and to ignore distractions. Mozart was the single most important external factor in allowing me to do the kind of writing that I needed to do."

An estimated 10 to 15 percent of young males in the United States—as well as a number of girls—are diagnosed with some form of attention deficit disorder (ADD) or attention deficit hyperactivity disorder (ADHD). Some of the symptoms of these disorders include emotional oversensitivity, anxiety, an inability to concentrate, difficulty relating to peers, and a general incoherence of thought. Though scientists disagree on the exact causes of ADD and similar disabilities, an increasing body of research supports this student's contention that the highly organized music of Mozart can greatly benefit children and adults who suffer from them. This is welcome news to parents of the more than one million American children who take Ritalin, an amphetamine, for ADD every day.

In a study of nineteen children aged seven to seventeen with ADD or ADHD, researchers played recordings of such Mozart compositions as *Eine Kleine Nachtmusik,* Piano Concerto No. 21 in C Major (K. 467), and *The Marriage of Figaro* during thrice-weekly neurofeedback sessions for some of the children. Nothing was played for the others. The researchers reported that those who listened to Mozart reduced their theta brain waves in exact rhythm to the underlying beat of the music, and displayed improved focus, mood control, and social skills. Seventy percent of the subjects who improved maintained that improvement for at least six months afterward.

The Tomatis Method, in which Mozart's music is filtered to remove the lower frequencies and played over headphones to children with learning disorders, is also helping an increasing number of patients through listening centers in Bethesda, Maryland; Seattle, Washington; Amherst, Massachusetts; New Orleans; Dallas; Phoenix; Toronto; Mexico City; and throughout California. As Sound Listening Center's Billie Thompson pointed out earlier, children are encouraged to draw, work with clay, or move about as they listen; their physical actions seem to accelerate the benefits of the music. This listening exercise gradually retrains the ear, easing the symptoms of ADD as well as certain emotional problems and physical challenges. "Mozart is an ideal transition from a world of noise to an orderly and organized thinking process," Thompson adds.

Approximately seven million children in the United States suffer from language-based learning disabilities (LLD), or reading problems. Many experts believe that LLD stems from a child's inability to distinguish among short, staccato sounds, such as *d* and *b*—a neurological impairment perhaps caused by chronic ear infections during infancy. Researchers at the University of California in San Francisco used sound intervention to treat children with this condition, drilling a group of five- to ten-year-olds three hours per day with computer-produced sound that draws out short consonants, like a record played too slow. Four weeks later, the LLD children who were one to three years behind in language ability had improved by two full years. The improvement has lasted.

If your child is diagnosed with any kind of learning disability, have her hearing tested before accepting the diagnosis as complete. Hearing loss is not the only possibility; some children experience a *hypersensitivity* to sound that prevents them from filtering or focusing on it, and that leaves them uncomfortably overstimulated. Hearing tests should reveal whether either of these hearing disabilities is causing your child's attention deficit, but be sure she is tested for auditory *processing* problems as well as auditory acuity.

MUSIC FOR THE MIND

The melding of mental and sensory memories is vital for productive learning, and actively making music combines these processes more effectively than simply listening to it. In fact, learning actively—that is, through the body—actually stimulates a different memory system in the brain than passive, or verbal, learning and involves more parts of the cerebral cortex. Active learning also produces stronger long-term memory. As I have pointed out, studies show that schools that include music in their curricula produce the highest achievers in the country. A study in Hong Kong found that adults who had received music training before age twelve have a better memory for spoken words than those who did not. Dr. Gordon Shaw of the University of California at Irvine has shown that taking piano lessons and solving math puzzles on a computer significantly improves specific math skills of elementary school children. A study of 7,500 university students in the 1980s demonstrated that music and music education majors had the highest reading scores of any students on campus, including those in English, biology, chemistry, and mathematics. The College Entrance Examination Board reported in 1998 that students with experience in musical performance scored fifty-two points higher on the verbal part of the SAT and thirty-six points higher on the math section, for a total of eighty-eight points higher than the national average. Research at New York's Beth Israel Hospital and elsewhere has demonstrated that musicians' brains are literally different from the brains of nonmusicians, possibly due to their daily experience of combined mental and sensory learning.

Are these the reasons why your six- to eight-year-old should consider learning to play an instrument, if she hasn't already done so? Not necessarily. Her own interest is paramount. Even if she doesn't want to take recorder or piano lessons, there are many other ways she can use music to enhance her learning skills, such as singing in the school chorus or attending dance class, and pushing her to play an instrument will only lead to resentment. Another reason to learn to play an instrument is for the sheer pleasure and beauty of the music. Beyond this, music making's ability to nurture creativity and self-expression, its power as a stress reducer, its links to your child's cultural and musical heritage, and the emotional resonance it creates within your child may all be more valuable than high scores on the SATs. As Lois Birkenshaw-Fleming points out, other subjects such as math and science may give children the tools for living, but music and the other arts are what give them a reason for living. This is reason enough for your child—*if she wants to*—to pick up an instrument and play.

A MOZART MUSICAL MENU

- Variations from the Sinfonia (K. 297b). Repetition is one of the best ways to learn information. When we repeat information or musical melodies in a variety of ways, the brain and the ear begin to listen differently. Ask your child to listen to this piece and see if she can tell what melody is repeated all the way through the piece. If she feels drowsy, invite her to stretch and gently move to this music to help the mind and the body prepare itself for study and concentration.
- Andante from the Symphony No. 6 (K. 43). Drawing pictures can help activate the right brain's natural spatial development. Offer your child paper and chalk or crayons, and allow the sweeping sounds of this Andante from Symphony No. 6 to inspire her imagination.
- Andantino grazioso from the Symphony No. 18 (K. 130). When your child feels the stress of study or exercise, this slow, magical piece can help her relax. Put on the CD, suggest that she close her eyes and let Mozart give her ears a gentle, relaxing massage.
- Andantino from the Symphony No. 24 (K. 182). This andantino is one of the most perfect pieces that Mozart ever wrote for bringing language to music. Ask your child to close her eyes as she listens to this piece, and see if she can hear the story it's telling. The strings and the flute are very clearly repeating the sound information. The violins and the flute speak in a clear, direct form. By listening to the ways in which music develops an idea, asks questions, and then allows the listener to come to

resolution, your child is introduced to a structure she will later use in writing essays and speeches.

- Allegro aperto from the Violin Concerto No. 5 (K. 219). This sparkling allegro can help charge the brain and give your child's ears a good deal of exercise. Have her spend ten minutes before her study time imagining that she is conducting the orchestra. This process of active listening and moving will bring her brain and body to full attention, preparing her for better concentration.
- Prestissimo from the Serenade (K. 203). This prestissimo, which means very fast, can allow your child to visualize her own dramatic production, including many characters with many different feelings. Suggest that she listen to this piece many times with her eyes closed, imagining a ballet, seeing what happens to the characters, and letting her mind and body move together.

CHAPTER 9

MOZART JR.

Creating a Sense of Identity (Eight to Ten Years)

Just as my fingers on these keys
Make music, so the self-same sounds
On my spirit make a music, too.
—WALLACE STEVENS

It has been many years since Mozart's variations on "Twinkle Twinkle, Little Star" first invited your child into a world of experimentation, variation, and richness. By testing his boundaries, fitting known concepts into new situations, and learning to greet novelty with delight instead of fear, your child has grown from a mysterious being silently listening inside his mother's womb to the vibrant, ever-surprising eight-year-old you know today. Now, as his communication skills, physical abilities, and mental development mature, he can embark on the process of integration—a remarkable feat and the essence of healthy growth.

For the eight- to ten-year-old child, maturation involves more than the physical, emotional, and intellectual experimentation of early childhood. Rather, it develops from a more complex process of encountering new information, examining its implications and potential against the background of one's previous experience, and then integrating that *pro-*

cessed information into a more complex and balanced whole. Educational theorists from Plato to Piaget have described this sequence as the way in which all older humans most naturally and efficiently learn. Interestingly, one of the foremost types of music that accompanies a child through the early years—the *sonata allegro* form epitomized by the simple sonatinas that every beginning piano student learns—perfectly mirrors and expresses this more mature process.

Mozart made frequent use of the sonata allegro form, whose three-part sequence of exposition, development, and recapitulation provided the perfect architecture for his playful yet revelatory style. Listen to a recording of one of his sonata compositions (such as the Piano Sonata in C Major [K. 545]), and you will hear how it begins with the *exposition* of his theme, a simple and recognizable melody that represents the statement of the music. After the theme is firmly established, the music moves into the *development* stage, in which the theme is altered in key, mood, melodic detail, background, or sound "colors." Finally, the sonata *recapitulates* the beginning melody, reviewing all the variations and revelations first presented in the development phase. The theme that began as a simple statement becomes ornate and full. The theme's statement is now stronger, enriched by new textures and harmonies, new tones and instruments. The journey is completed.

In these years before the turmoil of adolescence, great academic, social, emotional, and physical challenges await your child, and he will have to meet them at a deeper, perhaps more serious level than before. Yet Mozart's sonatas are there to remind him, from far away across the centuries, that listening to new things doesn't have to be difficult, really. It can be beautiful, intriguing, and even fun.

AN ORCHESTRATED MIND

If we ever do experience a personal age of reason, ages eight through ten are the most likely time for it to begin. After the rapid brain growth of the seventh year, during which the skull actually expands, your child is able to nestle into a period of increased mental depth, or neural integration, during which reading and mathematical abilities begin to mature, increased physical awareness and even poise are achieved, and a new level of self-expression becomes possible.

But all is not quiet inside that little head of his. Your child's brain is now gearing up for a two-year growth spurt in the auditory area, a region that has remained virtually unchanged for the past four years. This growth, between ages nine and eleven, will greatly enhance his ability to make fine discriminations in hearing and producing sounds and to recognize and reproduce nuances of inflection and tone. Such development, following the neural integration process, makes this a golden period for the arts. For the first time, your child will be able to seriously improve his musical, artistic, athletic, movement, and creative writing skills, because he will sense the contrast between where he is and where he wants to go. His increased attention span will help him stick to a creative task until he reaches his goal and will lend him a new physical grace or stage presence. He experiences an improved ability to blend with a group in choral singing, the school band, and athletic teams, and a new desire to understand the cultures and beliefs of others. Better listening also enables him to make great leaps in language acquisition—including foreign languages. This is a perfect time to provide him with plenty of vocabulary words and teach him the rules of grammar. It is also an excellent time for him to begin formal foreign language studies—though the language need only be introduced now through hearing, singing, and role playing, and not fully mastered. In fact, I call this the state of the arts age. It is a prime time to develop the tools and rules on which a lifetime of creativity will rest.

As the corpus callosum, the band of fibers connecting the right and left hemispheres of the brain, finally becomes insulated and fully functional during this period, the lateralization of the brain becomes complete. An increased ability to handle abstract ideas and logic lies ahead at ages eleven and twelve, and an increased sense of global vision or idealism in the early twenties, but in general, the drama of the brain's evolution gives way after this period to equally dramatic changes in the adolescent's body. The songs, rhythms, and patterns of early childhood have reached their *recapitulation* stage, a time when all the variety of your child's experience will inform his development and self-expression. As a parent, you are in for a delightful treat.

IN THE KEY OF ME: EXPLORING THE SELF

As your child moves from early toward middle childhood, his self-concept, his opinions and feelings about himself, grow correspondingly complex. The often highly positive, concrete descriptions of the early years ("I am a good runner," "I can ride a trike fast") are folded into more general beliefs ("I'm pretty popular," "I'm very smart"). These new assumptions are an important, perhaps decisive, factor in your child's motivation to succeed. Music, along with his increasing ability to express himself in artistic and creative ways, can help him explore such concepts as how he rates compared to others and to his younger self, where his talents lie, how well he relates to others, and even whether he learns best in visual, aural, musical, physical, or other ways. If he and you find that his weaknesses are generally social ones, music can lead him to new friendships and ways of relating to others. If they are academic, rhythm and tone can help him advance in his schoolwork and even enjoy his homework. If he feels less handsome or coordinated than other kids his age, musical exercises can help him gain a sense of grace and mastery. If he has had to deal with death, divorce, or other traumas, music can help him express his emotions and begin to connect him to others in a healthy way.

One way in which your elementary school child begins the long process of getting to know himself as an individual is by paying attention to his newly emerging inner voice. This voice, ideally nurtured from years of listening to music, to the voices of others, to tape recordings, and to his own vocalizations, will accompany him for the rest of his life. It will serve as his conscience in every ethical quandary; as his supportive friend in challenging situations; as his reality check when he encounters an intriguing but outlandish idea; and as his ally when the chips are down. The more clearly your child can hear this inner voice, the more attuned he will be to his true self—and the more creative, productive, and happy he is likely to become. One way of helping him make this connection with his inner self is to focus his attention on how he expresses himself outwardly. By experimenting with tone he will learn, as you may have learned while you were pregnant, that the quality of his voice and breathing can affect the way he thinks and feels. Recognizing this will encourage him to experiment.

TUNE IN, TUNE UP

TONING UP THE VOICE, TUNING UP THE EAR

Learning to recognize and rely on one's inner voice isn't an automatic process. One has to actively listen for it. One way your child can train for this task is by tuning up his ears, that is, listening to another person's pitch and trying to match it with his voice. By doing this, your child is exploring his own unique sound and strengthening the relationship between his voice and ear.

The ability to match pitch comes with practice, so don't expect your child to be instantly successful at this age. Certainly, there is no right or wrong tone; your child is just exploring. By practicing together a few minutes each day, as you wash dishes together or fold the laundry, you create a safe and fun way for two exploring voices and ears to come into harmony.

One way to help your child tune up is to sound one smooth tone toward one of his ears, and then let him find the tone with his own voice. If he misses the tone, don't say "No, that's wrong." Just say, "Now, let's listen on the other side." As you explore how each side listens to higher and lower pitches (it can sometimes help to make a megaphone with your hands and funnel the sound more directly toward his ear), observe whether one side is more naturally tuned than the other. When he matches the tone, raise your hand to signal the fact that you have achieved one voice. Continue the tone for a bit so that the two of you can enjoy the sensation. Then have him sound different pitches, and try to match his pitch in the same way.

When tuning up together, remember to:

- Use the voice to tune up the voice. A piano or guitar does not have the same tonal quality and can be confusing to the ear.
- Use an "oo" or "oh" sound in the beginning. In later stages, you can proceed to other vowel sounds, such as "eh" and then "ee."
- Find the most comfortable range for your child's voice. At this stage you should not be concerned with singing as much as teaching him to listen closely and to feel comfortable with his voice.

Every voice is different, and your child doesn't need to sing perfectly in tune to be able to listen and participate in musical experiences. No matter how off-key he may sound, be sure you remain supportive. His budding sense of identity is still very fragile, and criticizing his voice at this age is like criticizing his inner self. Remember—any voice can eventually be tuned, and his singing ability is not the most important concern at this young age.

Your child's exciting encounters with himself will inevitably take him into the realm of the emotions. If he has been actively listening to music since he was young, he will already be receptive to the wealth of emotional nuance around him. Certainly, this is the case with music itself: in a recent study of people aged six years to adulthood, the only groups who tended to break down music according to its expressive components, rather than analytical ones, were professional concert performers and untrained people with a strong love of classical music. To keep your child's heart and mind open to the emotional vibrations that exist in social relationships, in books, and in his own developing self—and to increase his awareness of all forms of emotional, social, and intellectual nuance—continue to expose him to the complex, expertly modulated sounds of classical music such as Bach's *Jig Fugue,* Chopin's *Raindrop Prelude,* Beethoven's *Moonlight Sonata,* and music from Debussy's *Children's Corner Suite.* Whereas in the past you led him through his listening experiences, it's now time to hand the baton over to him. Allow him to express to you, or to himself, or in writing, how the music makes him feel.

You might even suggest that your child literally take baton in hand, imagining that he is the leader of a great symphony orchestra as he conducts *The Nutcracker Suite, The Sorcerer's Apprentice,* or any waltz, march, or symphony with a strong, steady beat. If he has never seen a live symphony performance, it might help to rent a video of, say, Leonard Bernstein conducting. As he conducts, your child might wave his hands as he listens to exciting music, put one index finger to his lips as he listens to quiet music, reach high and keep the beat with his feet, fingertips, and then his elbows, or even close his eyes and be very still. Such an activity encourages him to express a variety of dynamic changes in sound through his body. Remind him that a conductor doesn't dance—he listens, anticipates moods, and clearly reflects a strong beat and feelings to the orchestra. Help him experience the process of sculpting feelings and beauty in the air.

At this age, your child can also begin to express his growing awareness of interpretive meaning through his own music making. If he plays an instrument, allow him the opportunity to hear how professional musicians have interpreted a piece he knows, and encourage him to imitate their emotional expressiveness. This is also a very good time to play your own favorite music for him, describing the circumstances under which it was written (if you know them), and telling him what the music means to you. Remember, including music other than classical selections adds dimension for both of you.

During this period of growth, your child may realize that he approaches life and sound differently from his peers. In fact, every child has his own way of apprehending the world—a style as unique as a fingerprint. Rather than expressing his emotions in words, for example, he may feel more comfortable expressing them in sound, in pictures, in movement, or in sports. He may also begin to notice, or you may notice, that he retains information much more effectively if it's in written form, or sung, or coupled with some kind of physical movement or human touch, rather than simply spoken. Cognitive psychologist Howard Gardner of Harvard University broke new ground in this area when he determined that human beings operate according to a number of different modes of intelligence. Each individual displays a unique balance of strengths and weaknesses among these frames of intelligence, and it is futile to try to categorize people by one learning style alone. The seven kinds of intelligence he originally identified include *linguistic, logical-mathematical, spatial, bodily-kinesthetic, musical, interpersonal,* and *intra-*

personal intelligence. (In 1995, Gardner added *naturalist* intelligence, denoting skill in making consequential distinctions in the natural world.) As we have seen throughout this book, music can help your child enhance his abilities in each of these areas. Now is a good time for him and you to take stock of precisely where his strengths and weaknesses lie, so that you can bolster his skills where he needs it and help him learn and grow in ways that are natural and productive for him.

SPOTLIGHT ON THE SPECIALIST

DAVID LAZEAR: MULTIPLE INTELLIGENCE

For the past decade, David Lazear has been among the foremost educators creating practical applications relating to Gardner's theory of Multiple Intelligence. Lazear's seven books on curriculum development, assessment, and creativity in the classroom (including *Seven Ways of Knowing, Seven Ways of Teaching,* and *Seven Pathways of Learning*) have given teachers nationwide the tools for applying Multiple Intelligence theories to the classroom.

Although he is not a music specialist, Lazear considers music's importance to reach far beyond simple music literacy and aptitude. "The auditory and rhythmic intelligence that music contains teaches us language, movement, communication, emotions, and visual/spatial intelligence," he tells us. "This vibrational, primordial way of knowing is the basis of music and actually is essential to the whole brain's function."

By connecting sound, movement, speech, and interaction with a musical component, it is possible to activate and integrate more of the brain than with any other educational tool. By drawing to music, speaking in different accents (the musical quality of language), rapping spontaneously, and becoming aware of both the active (playing an instrument or singing) and passive (listening, imaging, or using music in the background) aspects of music, children can improve their mathematics, language, coordination, social, and personal skills. The use of multiple forms of intelligence allows them to integrate and harmonize as well as use their brains to their greatest potential.

As your eight- to ten-year-old continues to explore all the nooks and crannies of his personality, you may hear him vocalizing in the shower or in his room much more often. He may also lip-synch a song in front of the mirror now and then, or dance in front of the full-length mirror when he thinks no one is looking. All of these are healthy and necessary steps in getting to know oneself, and the more he engages in them the fuller his experience of himself is likely to be. During this period, encourage him to create songs, music, and dance based on his own experience as much as possible, as well as to act out his favorite popular music. In this way, he'll discover his true expressive style. If he offers to share these insights with you via a performance for the family one evening, by all means get everyone together, serve refreshments, and applaud!

Greater neural integration means a greater appreciation for a fully structured story, and stories' enormous potential to show us new aspects of our own experience has strong appeal for children this age. This is a perfect time to expose your child to *program music*; that is, music especially written to depict a story. Two of the best-known examples of this genre are *The Sorcerer's Apprentice* and *Peter and the Wolf*. Others include *The Grand Canyon Suite* by Ferde Grofé, which magically allows your child to hear himself wandering through the landscape and surviving a thunderstorm. *The Moldau* by Smetana brilliantly evokes the story of a river in Czechoslovakia, starting with two little trickles (two clarinets), increasing in size as it flows through towns and villages, offering vignettes of life in the countryside before it joins the biggest river in the country. Your child can use nonclassical music to envision a story as well. Soundtracks to children's films are popular and mood enhancing at this age, and bring back the magic of the movie itself if your child has seen it. Historical songs such as "The Battle Hymn of the Republic" also evoke the power of storytelling, as do some evocative foreign ballads and songs.

Naturally, your nine- or ten-year-old child will begin to form opinions on the kinds of music he prefers and those he doesn't. He will explore genres of music to which you've never introduced him. He may decide to take up a new musical instrument and become interested in composing songs on the computer, guitar, or piano. In other words, he will begin to define his own tastes. This experimentation is yet another means of exploring the self, and in general should be encouraged. In fact, a varied musical diet is good for your child's mind and body as well as his sense of adventure. Mozart is usually best for study and for organizing thought, but when your child is working on a creative project or is grappling with issues that don't lend themselves to simple, linear solutions, jazz might better stimulate new ideas. The music of Miles Davis, John Coltrane, or other jazz greats can set the stage for the highly creative theta consciousness, the brain wave state associated with artistic and spiritual insight.

Some rock, rap, and other music centered on the beat can keep older children focused within a chaotic, unpredictable environment. It creates structure and sharpens their ability to organize. In contrast, New Age or ambient music allows highly disciplined or overscheduled children to unwind and float freely.

Samba and Brazilian music belong to what I consider one of the healthiest and most accessible genres of contemporary music. Brazilian music, which fuses elements of Latino, Indian, African, and indigenous South American traditions, has the improvisational quality of jazz, but just enough sweetness and drive to keep the listener attentive. This music can make your child feel safe, soothed, and energized. Playing or drumming along with the music, either alone or with family or a group of friends, relieves tension and instills a wonderful sense of community.

By trying out different types of music, your child can learn how various genres affect his psyche. With freedom to choose his own influences (and no untoward pressure to give up the music he loves), he may well choose rock as an emotional outlet, but switch to Mozart before starting to study, even when he's a teen. By encouraging him to share his evolving musical passions with you, you are showing him how open you are to him in *all* aspects of his life. So ask him to explain to you what he loves about certain selections, and support his desire to attend live performances and to express his own emotions musically. You may be rewarded by having an open line of communication during those difficult years of adolescence.

THE WOLF GANG

Ages eight to ten were magical ones for Mozart. Having begun to create his own solo performance pieces, he moved on at age eight to conduct his first symphony at the court of King George III in England. This experience of public, ensemble music making must have meant quite a lot to him: relatively isolated as a child prodigy, he craved contact with others who thought as he did. This longing to join a group is typical of nearly all children of this age. As your child becomes more familiar with the boundaries and definitions of his own personality, he is increasingly drawn to the sense of being part of an amorphous crowd, partly for the challenge of retaining his own separateness within the group. The push-pull process between the need for independence and the longing to be part of the pack may explain a great deal of behavior during the final few years of elementary school. It also makes this an excellent time to encourage a joyful enthusiasm for life in community by joining athletic teams, the school band or orchestra, choirs or choruses while still allowing room for your child to be on his own.

Music educator Nick Page describes one music teacher's experience with slightly older children that illustrates the power of this desire to belong. The teacher, named Betty, taught in a middle school with a high dropout rate, low test scores, and a number of drug-related problems. To help her students deal with these challenges, she designed a group drumming program for the most troubled kids. Using West African percussion instruments, shakers, bells, sticks, and drums of different sizes, she taught the group a series of rhythmic patterns, gradually layering one part on top of another. After that, she let the rhythms take over. "For twenty or thirty minutes the students would explode with these patterns," Page tells us. "Little improvisation was allowed, as Betty was teaching the 'classical' West African art of drumming, where only the master drummer is allowed to improvise, and then only with strict rules." But the students seemed to revel in the knowledge that they were playing the same parts that their original African ancestors had played.

Since the drumming group began, the teacher reports, the participating students have become more serious in their work, more attentive in class, and more confident of their abilities. Perhaps their success at sustaining a pulse increased their ability to sustain their attention. Perhaps the highly charged entraining energy of the percussion instruments enhanced their brain activity. Certainly, the drumming helped shape a positive identity for the entire group. Such sessions inspire a powerful sense of belonging, not only to a cohesive group but also to an ancient but living tradition. By forming a bond with the musicians of a different culture and a different time, the drummers of West Africa, these children were able to see beyond the bleakness of their everyday lives. The group energy they experienced gave them a boost, and for a few moments they were able to see over the treetops and realize that there is much worth living and working for.

But there is more to group music making, or even listening to music—something that seems to bestow a group serenity or transcendence and banish all fear and anxiety, if only for a moment. Most of us have experienced this feeling during a concert or when singing hymns at church; leaving the symphony hall or church, we feel refreshed and inspired. Carolyn Dondero, a certified reading specialist in California, wondered if she could take advantage of this musical mood stabilizer in her work with ninth- and tenth-grade children suffering from lapses in emotional or neurological development, students who displayed plenty of adolescent angst and defiance into the bargain. Curious about how music might alter the classroom atmosphere, she began playing classical recordings softly during the six minutes it took for the kids to enter the room, get out their books and supplies, and so on. "A most heartening result of using music," she writes, "is that the classes each demonstrated more caring than ever before. An example occurred as I said good-bye to the students. Sometimes in the past I have sent them off, saying, 'Aloha, ciao, adios, sayonara,' waving and smiling, and being ignored. This past year, more students responded with, 'Take care,' 'Bye,' 'Have a nice weekend,' 'See ya,' than ever before. The students became unusually helpful at taking care of the room by not putting garbage on the floor. Many students exhibited spontaneous kindness. The tone of the classes was often warm and friendly. Many days, both with and without music, had no 'off task' events."

Jo Ann Reeves, a fifth-grade teacher of high-risk children in Austin, Texas, has developed her own form of communal music to encourage a new mood and a better self-image. As the children take their seats in a circle at the beginning of the school day, she sits in a rocking chair, rocking to the rhythm of slow baroque music that plays softly in the background. By remaining in this listening position, she communicates to her students that she is ready to hear them express their joys and concerns. The students do this by passing a drum around. Whoever has the drum gets to speak. Reeves reports that the students laugh, cry, and bond into a team of better-motivated, more optimistic learners. "The day starts with a tune-up. We stay in better harmony all day."

You don't have to be a teacher to encourage your child to make use of group musical experience to increase his confidence and self-esteem. This is a good time for him to engage in circle dances, square dancing, ballroom dancing, or any kind of singing or music-making group. As your child moves from his eighth birthday to his ninth and then his tenth, keep an eye out for such programs in your community, and help him get involved.

Of course, your child will have trouble participating fully in a group's magic if his verbal communication skills are below par. Keep in mind that poor speaking ability is usually a result of poor listening skills. You may have helped your child listen actively for a number of years already, but as his brain's auditory centers develop (and as he approaches adolescence), the quality of his listening becomes even more important. Paul Madaule, director of the Listening Centre in Toronto, Canada, has described his own experience as an adolescent who did not listen or speak well. "I had the kind of voice no one cared to listen to—not even myself, as I later came to realize," he writes. By practicing a series of listening exercises recommended by Dr. Tomatis, Madaule not only learned to speak effectively in group situations but also realized that "the evolution of my use of voice, speech, and language went hand in hand with an increase in my level of energy. As a teenager, I was always tired. I could spend entire days doing absolutely nothing while my mind was running a hundred miles a minute. I always had grand plans that never saw the light of day. My energy was blocked. Now, with my busy schedule of long workdays and no weekends, I often get little sleep for weeks at a time. The energy flow is not blocked anymore; it is there when I need it."

Madaule was so inspired by the way that improved listening completely changed his life that he went on to do research in listening and the brain, and to direct the Listening Center (now in its twentieth year). Meanwhile, he developed a series of listening exercises, called "Earobics," based on Dr. Tomatis's work. One of these, a reading exercise that Madaule calls the most complete and the most efficient of all the Earobics, is described here.

A MUSICAL RECIPE

PAUL MADAULE'S "EAROBICS 10"

Have your child sit close to a table or a desk in the "listening posture." This means sitting erect but relaxed on a stool, inhaling, and relaxing from the toes upward until his shoulders feel light. Next, have your child hold his right fist to his mouth like a microphone, with his right elbow resting on the table. Now tell him to begin reading aloud. While reading, he should imagine himself telling a story to a group of people. He has to captivate them. By pretending to tell, instead of read, the story, his voice will probably project a richer quality, clarity, and strength. It doesn't matter if your child makes mistakes, as long as he makes them out loud, with a strong, clear voice. He needs to listen to himself make mistakes in order for his brain to distinguish the difference among the sounds he emits and the letters he reads and trigger the correction.

It wouldn't hurt for your child to read aloud for fifteen to thirty minutes a day as long as he is in school. You can make this more enjoyable by acting as his audience, responding with interest and enthusiasm as his voice becomes increasingly smooth and rich. Not only does this exercise help with verbal expression, but it is most helpful when your child is tired or depressed, when his brain is fogged and he is having a hard time pulling himself and his ideas together. Reading aloud is also an efficient way to reduce tension headaches. It helps us feel light, awake, vigilant, on top of things, and to think clearly and in an organized fashion. Consider these benefits as you sit down to read aloud to your child at bedtime tonight!

Just as traveling abroad makes us love home even more, older children feel a natural impulse to return home for reassurance after having explored the larger world. The wise parent welcomes a son's or daughter's desire to touch base before venturing into the grand world of middle school. You can fulfill your child's need to feel the support and cohesiveness of his immediate family, while still hearing his own, unique voice, by making music together, just like in the old days when he was a toddler. At this age, your child will especially relish multipart songs casually attempted while, say, driving in the car. Try singing "Swing Low, Sweet Chariot" at the same time your child sings "All Night, All Day." If there are other family members present, some can sing one song and some the other. The two songs meld perfectly, require enough concentration to challenge both of you, and sound so good together that your child will deem them "cool."

Family relationships are also enhanced by creating your own family orchestra, with one person singing, another on piano, and another perhaps drumming a rhythm on the coffee table. You can display and get to know new sides of one another by inventing funny songs about one another, creating story songs about family experiences, or even having a singing dinner one night instead of ordinary conversation.

By the third grade, children start to become more aware of cultures and styles other than those they grew up with. While starting to learn a new language is an excellent activity for a child this age, you can also help him expand his knowledge of other ways of thinking by continuing to expose him to music of other countries, talking to him about the cultures within which the music was formed, and encouraging him to learn to sing Spanish, French, and other simple foreign songs himself.

You can also help your child increase his social intelligence by helping him make musical contact with other cultures and groups in his own area. Take him to concerts, musical dramas, and other such performances aimed at groups different from his own. If he plays an instrument, look into the idea of having him mentor a younger child, volunteer to help teach music groups in other parts of the region, or even help children with special needs learn music. He will quickly learn that music acts as a bridge linking all people, no matter who they are. Through his developing love of music, he can express his love for, and interest in, his entire community.

MUSIC: THE GREAT MULTIPLIER

"Several years ago, my youngest son, Dustin, was having trouble learning his multiplication tables in school," writes music educator Kerry Hart. "After two years of struggling, with many conferences and threats of retention from his fourth- and fifth-grade teachers, the only suggestion Dustin's teachers could give my wife and me was to 'drill' him every evening. I knew this wouldn't work, because Dustin is one of those right-brain children who has had trouble responding to the traditional verbal-linguistic (lecture) approach. I began thinking of the realms of intellectual functioning in which Dustin excelled: the visual-spatial, physical-kinesthetic, and musical-rhythmic. Without giving much thought or rationale for the action I took next, I sat down at the piano one evening and began playing the old familiar children's song 'Are You Sleeping.' I asked Dustin to sing along and changed the words to the multiplication tables as we sang. In one evening, he had memorized the multiplication tables through six; and a week later, we took the same approach with the rest. He memorized these in one evening as well and didn't have a problem remembering over the years.

"About a year after the multiplication experiment, Dustin came home from school and commented that he had failed a biology test over the order of classification. He mentioned that his teacher was going to retest the class the next day, because most of the class had failed. Using our pet cat, Unis, as an example, I quickly changed the words of the song "Reuben, Reuben, I've Been Thinking" (a song that came to mind that seemed to fit with the rhyme of the order of classification) to show how a cat fits into each classification. He received an A on his test the next day, and now, years later, still remembers the order of classification by singing the song."

Hart was inspired to create a series of songs to teach some of the most difficult and important concepts taught in grades one through five, in the areas of language arts, mathematics, science, and social studies. Naturally, he used Dustin to test the effectiveness of his approach. When reviewing concepts through music, Dustin's test scores improved by an average of 20 percent.

As schoolwork becomes more challenging in the third grade and later, it pays to consider the ways in which music can support your child academically. One clear way it performs this role is by altering your child's mood. Over the years, you may already have used music to move your child gently from one activity to another, to modulate his activity level or alertness to better match the task at hand. Music can continue to set the mood for important activities as he begins his serious career as a student.

Robert Cutietta, interim associate director of the Saint Paul Chamber Orchestra's Neighborhood Network of Education, Curriculum and Teachers (CONNECT), has helped schools in the Saint Paul area use music in their curricula since 1994. Music is integrated into the schools' nonacademic subjects, and students attend performances given by the orchestra. In a review of the program conducted in the 1996–1997 school year, Cutietta reports that the third graders in the program reported significantly more positive attitudes than previously toward the school staff and while doing schoolwork. They also reported less friction in their classes than did students whose schools weren't in the program. This is important because the overall social climate in the classroom has been identified as an important aspect of academic achievement. Studies have shown that a student's perception of his classroom environment is correlated to his success in mathematics, social studies, science, and numerous other areas.

Studies have also shown that the home environment plays an equal, if not greater, role in motivating students to excel. One longitudinal study following children through early adolescence showed that the home environment at age eight has a statistically positive and significant effect on the child's academic motivation all the way through early adolescence. The children's socioeconomic status did not affect these outcomes. Clearly, the more you continue to stimulate your child's mind and heart with music and the more you prepare him for study through the sounds in his environment, the better he is likely to perform as a student in the challenging years to come.

RHYTHMS OF STUDY

Schoolwork becomes substantially more challenging in third grade, and the amount of homework usually increases. Getting to know his personal rhythms, style, and energy level will help your child plan for and enhance his homework sessions. If he tends to come home tired at the end of the day, he might benefit from a brief nap or ten or fifteen minutes of some lively Mozart tunes, singing aloud, or even a brisk walk before settling down to start his homework. You might even offer to do a little deep breathing with him, or to read him a story while playing a beautiful, romantic music selection, such as Beethoven's *Pastoral Symphony*, the Symphony No. 6, in the background.

When helping him with homework, reading, or any other focused activity, watch his body language for signs that he's beginning to tire. Is his breathing shallow? Are his shoulders rolled forward? Is he fidgeting? Does his facial expression say, "I'm confused"? Does the tone of his voice tell you he's not interested? Is he looking away? If so, then it's time to stop or change the pace. Ask him to stand up, and lead him in taking a two-minute *aaaah* break, starting with a barely audible sigh and increasing the length, volume, and range of pitch with each sigh. It's best to start at a slightly higher pitch each time, and end with a much lower one. Or else, do a body "blither." Start by vigorously wiggling your hands, then add shoulders, hips, legs, and feet, and finally your head—all while making the "bllbllbllbll" sound that occurs when you relax your mouth and cheeks while vigorously shaking your head. A "hip, hip, hooray for school" (or for learning, or for Friday) is also appropriate, as is making several sirenlike noises, sounding a high pitch, sweeping it downward as low as you can go, and then bringing it back up again. Or you might simply play two minutes of march music while your child stands with closed eyes and imagines, without moving, marching through his learning tasks with great enthusiasm and success. In this last exercise, your voice can create excitement as you lead the imaginary march through various subjects and activities. Waltz music can be used to imagine dancing through the day as well.

These kinds of breaks will allow your child to clear his mind, calm down, and focus more effectively on his homework. If he still has trouble concentrating (or keeping himself awake), suggest that he change positions, move to another spot, or do his work out loud. He might even alternate voices to give his body an internal workout—doing his math in the voice of a very old man, and his reading as a very young child. The latter technique is especially helpful for memorization. The unusual sounds will anchor the information in his memory.

An important part of helping your child create good study skills is teaching him the importance of rewarding himself for a job well done. For example, you might help him be "composer for the day," creating a melody or rhythmic pattern preferably with bells or a piano. Write down the melody and lyrics, if there are any, either in standard musical notation or the simpler form I described in chapter 6. He may prefer to create a story instead, or a dance to his favorite music. In the evening, when everyone is at home, help him perform his composition for the family.

As your child grows to know himself better over the years, he will come up with his own ways to increase his focus and enhance his study skills. Avoid being overly rigid about the kinds of music or break activities he uses. Different music may be appropriate at different times.

Music's sensory power, its emotional richness, and its ability to speak to both the left and right sides of the brain make it an excellent tool to capture your child's attention and align his thinking before, during, and after study. As he grows in independence, encourage him to experiment with new ways to use music in his efforts to get ahead. Whether or not he comes up with brilliant new uses for rhythm and tone, he will be focusing on the development of his academic skills, and he will come to associate music with both learning and fun.

A FULL AND LOVELY VOICE: THE PROCESS OF INTEGRATION

An acquaintance of mine, David, was in a car wreck recently. His sports car swerved on the highway, hit the metal barrier, swung back into traffic, and did a full turnabout before coming to rest facing the wrong way in the center lane. Fortunately and miraculously no one was hurt. But the experience left David very badly shaken, and he felt a need to describe it to me in every last detail over dinner the next night.

As he came to the conclusion of his story, David set his after-dinner cup of coffee down in its saucer and looked at me incredulously. "Do you realize," he said, "it took me the entire dinner to describe to you what went through my mind during the accident—but the whole thing happened in about *five seconds*?"

Such is the miracle of multitracking, the powerful ability of the brain to process different types of information simultaneously. This parallel processing ability is largely responsible for the undeniable fact that the sum of the brain's energy (the mind) is greater than its parts, and is one reason why computer programmers find a true thinking computer such a challenge to design. As your child's brain begins to integrate more and more new information in increasingly complex ways, its multitracking abilities become more and more vital. Music's ability to mirror this parallel processing, as it affects so many neurological areas simultaneously, gives it a unique power to guide your child's mind to the next, more complex level.

Mozart's wife, Constanze, once commented that "Mozart wrote down music in the same way as he wrote letters." He often composed an entire movement in his head while playing billiards or walking about town. He would then write it down later with very few changes. In fact, at only age fourteen, the musician astonished the world by copying from memory a fifteen-minute piece composed of nine separate voices, music he had heard sung only once, at the Vatican to celebrate Easter.

Clearly, Mozart's was an extraordinarily integrated mind, capable of extremes of multitracking far beyond the capabilities of most. He was also hyperaware of music's equally amazing power to handle complexity. Neurologists often tell us that the typical human being can only think about approximately seven variables at a time. This may certainly be true when it comes to a grocery list or some other linear sequence. Yet an orchestra leader conducting a symphony—or my friend David driving his car—is clearly having many more than seven thoughts at a time. While it is true that complex thinking is possible in the mental realm alone, thinking that involves the body is infinitely more multidimensional. Moving a stick shift or turning a steering wheel can be taken care of through physical, procedural memory, and so more information can be processed in a given amount of time. In the same way, music's connections to the body as well as the brain give it enormous power in helping us to think in complex ways.

Music works to enhance our children's thinking not just through its complexity, but also through what many refer to as its integrative quality, its quality of truth. Ask a person how he knows that a statement is true, and he is likely to say that he feels it in his body, that he has a "gut feeling" or "knows it in his bones." These clichés reflect the fact that truth cannot be comprehended solely in the brain. The body must be involved, that subtle, physical response experienced, for a person to *know* that something is real.

In order to recognize that feeling, however, it is important to have felt it before. One way to introduce your child to this physical sense of coherence is to sing with him. It is amazing, when you think about it, that not only can two people match notes, but an entire roomful of people can also sing in one voice. That sense of communion, the same transcendent feeling that draws your child to such group experiences as drumming and music making and ball playing, is the same as the physical sense of correctness. Without it, full neurological integration is impossible.

Scientists tell us that the most complex, individual organisms in nature are also the most integrated ones, and we humans at our best are no exception. Our children's integration process began slowly from birth with the gradual development of inner speech. At first, whenever your child was thinking in words, you were likely to hear those words, because he would speak them aloud, as though addressing another person. At around age two, he probably began narrating his actions as he played, just as we often silently narrate our own activities as we pursue them. You may have even heard him announce "Don't empty the trash can!" even as his body irresistibly turned the can upside down. Now, between the ages of eight and ten, the voice is finally becoming internalized, integrating his perceptions and enabling him to better regulate his behavior, improve his reading skills, and attain a higher level of cognition.

A child whose inner voice is not fully in place will probably have problems with impulse control. He may need to move in order to think, and be unable to process information that is delivered to him in the form of classroom lectures. His attention span may be shorter than the norm, and he may need extra time to formulate answers to questions that are posed to him. Of course, this behavior is perfectly normal for children through about age eight. Gradually, though, if the inner voice is sufficiently nourished, these symptoms of immaturity steadily decrease.

Sadly, most schools today do not address the need for this development. The formal, non-movement-oriented, concrete presentation style so often used today is appropriate for a child who has already developed a reliable inner voice, but not particularly helpful to the child who hasn't. As we will see, movement classes such as Dalcroze Eurhythmics can do a great deal to meld mind and body. Practicing a musical instrument has a similar effect. However, it is important to show equal sensitivity to this developmental phase on the home front. A home filled with the constant noise of TV, computer games, and shouting people is not a good place for a child to learn to hear himself. Likewise, a child who is ordered impatiently to answer quickly while formulating his thoughts loses the chance to practice interacting with his inner voice. Giving a child time, giving him plenty of nurturing silence, and reading aloud to him in a quiet but expressive voice are all ways to lead him toward a serene, confident, and deeply integrated mind.

Dalcroze Eurhythmics, one of the many music education programs your school may offer, was developed in the 1890s by the Swiss music educator Emile Jaques-Dalcroze. Through classes he taught at the Conservatory in Geneva, Dalcroze worked to devise musical exercises that would develop more acute inner hearing as well as a neuromuscular feeling for music. His ear-training games were aimed at sharpening students' perception and creating a more sensitive response to the musical elements of performance: timing, articulation, tone, quality, phrase feeling.

Dalcroze also noticed that when listening to music, people tend to move in subtle, spontaneous ways that reflect the music's properties. Capitalizing on these natural gestures, he asked his students to walk and swing their arms or conduct as they sang or listened, responding to the movement or flow of the music with variations in time and energy. He called this study of music through movement eurhythmics, from the Greek roots *eu* and *rhythmos,* which means "good flow" or "good movement." He also encouraged his students to discover the music within themselves and to express themselves musically through keyboard improvisation, just as they might express an idea through speech, an emotion through gesture, or bring an image to life in a painting.

Dalcroze's students believed that his approach to music making caused them to hear, feel, and express music with their whole being. Music sharpened their aural and muscular sensations, and they responded with heightened mental, physical, and emotional consciousness. The classes created a bit of a scandal at the time, since Dalcroze insisted that the students dance in bare feet and bare legs. However, by the early 1900s children's classes had been created all over Switzerland, and in 1915, the New York Dalcroze School was founded. Today, Dalcroze techniques are found at every level of music education, in universities, public and private schools, and private studios.

The basic principles of Dalcroze Eurhythmics remain the same: students must develop an inner *aural* sense of music and an inner *muscular* sense. In this way students internalize the time, space, and energy relationships in movement that correspond to those in music. One exercise, for example, requires a child to learn to adjust his space and timing as he takes a walk (or a run) to music, attempting to arrive at a certain place by the end of the song. Older students develop an inner sense of time, space, and energy relationships by rolling a ball to the musical phrases. In this way, links are constantly being reinforced between the ear, eye, body, and mind. Watching a class, you may think your child is just stepping, clapping, gesturing, singing, and playing. But he is really well on his way toward developing an efficient, integrated brain.

FOR YOUR CHILD'S HEALTH

OUT OF SYNC AND OUT OF TUNE

You may have never heard the term "sensory-processing disorder," but we all remember the kids in our school who suffered from one. Though they may have been very intelligent, they often appeared disheveled, disorganized, and out of sync in both physical movement and social interaction. If you worry that your child lacks the physical and social integration that he needs, music may offer a solution.

Lois Hickman is an internationally known occupational

therapist as well as cofounder of Belle Curve Records, a company that produces audiotapes for children with special needs. She describes treating Mark, an eight-year-old who annoyed adults and most other children with his high, piercing shrieks, pushing, roughhousing, and other disruptive behavior. Mark was tested and found to be extremely sensitive to sound, touch, and visual stimulation. He had trouble with balance and gross motor skills, and he was unable to follow through on sequenced activities. It became clear to Hickman that Mark's kicking, hitting, and loud screeching noises were his way of achieving a sensory experience intense enough to give him a feeling of control.

Hickman addressed Mark's problems with shrieking by helping him explore different sounds and define how they felt as he made them. How did it feel, she asked him, to go from a high shriek to a deep, rumbling sound he could feel in his chest and stomach? Did the lower sound feel more like a "guy" sound? Did it make him feel stronger? Hickman played an audiotape for Mark of Native American drumming and chanting. Mark was fascinated by the vibrating sounds, music, and drumming and enthusiastically listened to and chanted along with the tape at every subsequent session.

Intrigued by Mark's fascination, Hickman began weaving the music into the other movement exercises that she had found helped the boy feel calmer and more focused: heavy work such as running a vacuum cleaner with the brake on, vibration (he loved to hold a battery-operated vibrating pen against his face), and climbing through tunnels and obstacle courses. As he worked, climbed, pushed, crawled, and otherwise explored the movements of his body, he chanted along with the tape and let its soothing rhythms enter his body. As his sessions continued, Mark's habitual tone of voice gradually grew less aggressive and more conversational, a change that led to many other positive changes at home and school. Clearly, music with a steady, vibrating drumbeat, and the chanting harmony of men's voices, had helped Mark feel the centering beat that could override his confusion and distractibility, and allow him to find new and better ways to express himself.

You can draw on Hickman's experience to guide your child toward greater physical, social, and psychological comfort. If you sense a certain awkwardness in his demeanor that he doesn't seem to be outgrowing, try the techniques described above. As always, listening, singing, and swaying or dancing to music together will help your child integrate music's rhythms and structure into his body. If he still seems to have difficulty integrating his body and mind, consider talking with a certified music therapist about other ways to help him now, before any real damage has been done.

THE WORLD BEYOND: GROWING UP WITH MUSIC

As a child moves through the second half of elementary school and toward middle school, the mind and body begin to change in ways that are less and less within parents' control. A separate self is emerging from the cozy nest of rhythm, tone, and melody, and a strong new ego begins to sense the need to express itself. The music and listening exercises you have performed with your child over the years have provided a very powerful push toward effective self-expression and a satisfying, creative life. Gradually, your child will discover that creativity means more than producing an artistic work. He will learn that he can be creative in relationships, in conversation, in academic work, and in practically every other aspect of his life. He will understand that beyond his love of quality sound, music has taught him how to communicate more effectively. It has introduced him to the people and the sounds of the world. It has framed his emotions and guided his body. Best of all, it has enhanced and informed the relationships within his family. You have taught him how to use music to celebrate joy, as well as to express less jubilant feelings. You have shown him that learning to listen in a variety of ways helps him feel more at ease, and that music, both the music he makes and the music he hears, can be a trusted friend throughout his life.

A MOZART MUSICAL MENU

- Allegro moderato from the Violin Concerto No. 2 (K. 211). This violin concerto has helped tens of thousands of children retune their ears at more than two hundred Tomatis-inspired Listening Centers throughout the world. Dr. Tomatis would suggest that your child listen to this piece with his right ear turned toward the stereo speaker, thus stimulating his language centers before studying.
- Andante from the Symphony No. 17 (K. 120). If he is already energized, this is a good piece for your child to listen to before studying. The high frequency and the slower tempo allow him to slow down and yet be stimulated at the same time.
- Adagio and *Gran Partita* from the Serenade No. 10 (K. 361). The *Gran Partita* is one of Mozart's most elegant and brilliant pieces. Have your child close his eyes and imagine he is in the royal courts of Salzburg or Vienna. As the music plays, he can watch the story unfold.
- Andante from the Symphony No. 15 (K. 124). This andante, which Mozart wrote when he was quite young, is a wonderful selection to begin your child's study session as he sits down and looks at the materials in front of him. What does he need to do? What does he need to read? How long does he have? While listening to this music, he can organize his desk, his ear, and his mind.

POSTLUDE

An Unfinished Symphony

When I was five
Margie Gahn said
I was not g r a c e f u l enough
to be in her dancing class
but I should WATCH.
I thought I was pretty good.

When I was nine
Miss Laughinghouse said
I was not t u n e f u l enough
to be in her singing class
and I must LISTEN.
I thought I was pretty good.

When I was forty-seven
Pearl Levine said
I was not e v o l v e d enough
to be in her consciousness raising group
but I might AUDIT.
I thought I was pretty good.

If they had said
come dance with us—
come sing with us—
come grow with us—

Ooooooooohhhhhh
What a difference!

—JUDITH MORLEY
from "Miss Laughinghouse and the Reluctant Mystic"

Every day of life is a new composition, based on themes that begin to develop before we are born, and on the patterns and melodies that reach our senses from our siblings, our parents, and the world around us. By wisely orchestrating the movement, gestures, sounds, and rhythms you encountered in this book, you created optimal patterns for your child to develop the unique form of creativity known as identity. In chapter 2, I described the African custom of singing a child into conception and birth, using song to introduce the infant to her family and her tribe. As your child passes through the one-way door to adolescence, she will take the melody you yourself created, reshaping it to suit her still-developing character and her greater world. The new tribal members who hear her song may live many thousands of miles away, belong to a vast array of cultures, and speak languages she has never even heard. Music will be there to build the connecting strands among them, for each person on the planet shares the early rhythms and rhymes they learned in childhood.

Author Robert Jourdain once wrote, "For a few moments music makes us larger than we really are, and the world more orderly than it really is. . . . As our brains are thrown into overdrive, we feel our very existence expand and realize that we can be more than we normally are, and that the world is more than it seems. That is cause enough for ecstasy."

May you and your children express that ecstasy through the human symphonies you have created together—the melodies of mind, the rhythms of the body, and the great musical creations of the self. Of course, children are nothing but surprises. There is no way to predict the outcome of a sonata, an opera, or a life. But by committing yourself to music, you have made a difference for your child. Someday the cycle will begin again. Your child will become a parent in turn, relishing the interchange among tone, rhythm, melody, and life experience, and passing on the family's legacy to children, grandchildren, and all the generations to come. It is my wish for all of you that your legacy include Wolfgang Amadeus Mozart's own words, "Love, love, love, that is the soul of genius."

SOUND RESOURCES

ORGANIZATIONS

American Association of Kodály Educators
1457 South 23rd Street
Fargo, ND 58103
(701) 235-0366
Offers training in the Kodály method.

The American Orff Schulwerk Association
P.O. Box 391089
Cleveland, OH 44139
(216) 543-5366
Offers training in the Orff approach at fifty-three locations in the United States and in many countries, featuring wooden xylophones, metal glockenspiels, as well as poems, rhymes, games, storytelling songs, and dances.

The Children's Group
1400 Bayly Street, Suite 7
Pickering, ON 11W 3R2
Canada
(905) 831-1995
(800) 757-8372
E-mail: moreinfo@childrensgroup.com
Web site: www.childrensgroup.com
Recordings, research, and other valuable links on the benefits of classical music.

Foundation for Music-Based Learning
P.O. Box 4274
Greensboro, NC 27404-4274
(910) 272-5303

Kindermusik International
P.O. Box 26575
Greensboro, NC 27415
(800) 628-5687
For information on local classes.

The Listening Centre, Tomatis Canada
Paul Madaule, Director
599 Markham Street
Toronto, ON M6G 2L7
Canada
(416) 588-4136
The oldest North American center and principal Canadian resource for the Tomatis Method. Call for information on all Tomatis centers in Canada.

Music Educators National Conference (MENC)
1806 Robert Fulton Drive
Reston, VA 22091
(703) 860-4000
Fax: (703) 860-1531
The largest association that addresses all aspects of music education, with more than sixty-five thousand members.

Music Together
Kenneth K. Guilmartin, Director
66 Witherspoon Street
Princeton, NJ 08542
(800) 728-2692
Web site: www.musictogether.com
For information on music and families, parent teacher training, and Music Together class locations nationwide.

Music for Young Children
39 Leacock Way
Kanata, ON K2K 1Y1
Canada
(613) 592-7565; fax (613) 592-9353
E-mail: myc@myc.com

Musikgarten
409 Blandwood Avenue
Greensboro, NC 27401
(800) 216-6864
E-mail: musgarten@aol.com

National Association for the Education of Young Children (NAEYC)
1509 16th Street, NW
Washington, D.C. 20036-1426
(202) 232-8777 or (800) 424-2460
Fax: (202) 328-1846
Web site: http://www.naeyc.org

National Guild of Community Schools of the Arts
40 North Van Brunt Street, Room 32
P.O. Box 8018
Englewood, NJ 07631
(201) 871-3337
Fax: (201) 871-7639
E-mail: almayadas@worldnet.att.net
Web site: http://www.natguild.org

Sound Listening & Learning Center, Tomatis USA
Billie Thompson, Ph.D., Director
2701 E. Camelback, Suite 205
Phoenix, AZ 85016
(602) 381-0086
A principal Tomatis center in the United States. Call for information on the dozen Tomatis centers throughout the United States.

The Suzuki Association of the Americas
1900 Folsom Street, No. 101
Boulder, CO 80302
(303) 444-0948
American headquarters for the Suzuki method and training.

SERVICES

High/Scope Educational Research Foundation: Resources for Educators 1999
600 N. River Street
Ypsilanti, MI 48198-2898
(734) 485-2000
Fax: (734) 485-0704
Order line: (800) 407-7377
A catalog of educational and musical movement resources for teachers.

Remo, Inc.
(805) 294-5600
Suppliers of a light, handheld synthetic drum useful for group or family drumming.

Suppliers of Musical Instruments for Children
MMB Music (800) 525-5134
Music in Motion (800) 445-0649
Music Is Elementary (800) 888-7502
Rhythm Band, Inc. (800) 424-4724
West Music (800) 397-9378
Small drums, egg-shaped shakers, jingle bells, keyboards—all of these companies provide sturdy, high-quality instruments to stimulate and inspire young children.

RECORDINGS

Audio-Therapy Innovations, Inc.
P.O. Box 550
Colorado Springs, CO 80901
(719) 473-0100
Recordings produced by pioneering infant-music producer Terry Woodford, designed to ease your baby's transition from the womb to the outside world.

Fearless Dentistry
Robert A. Wortzel, D.M.D.
1122 Route 22 West
Mountainside, NJ 07092
(908) 654-5151
Audiotapes and videotapes relating to young children and dentistry, focusing on helping young children conquer their fear of dental work.

Love Chords
The Children's Group (Babies)
1400 Bayly Street, No. 7
Pickering, ON L1W 3R2
Canada
(800) 668-0242
(800) 757-8372 (in Canada)
Dr. Thomas Verny's recording of soothing and stimulating classical selections for the pregnant mother and her developing child. CD notes include a twenty-four-page instructional booklet on sound exercises to use during pregnancy.

The Mozart Effect—Music for Babies, Music for Children, Music for Moms
The Children's Group
1400 Bayly Street, Suite 7
Pickering, ON L1W 3R2
Canada
(905) 831-1995
(800) 757-8372
E-mail: moreinfo@childrensgroup.com
Web site: www.childrensgroup.com

Music for the Mozart Effect, Vols. I–V, Music for Study, Music for Stress Reduction
Spring Hill Music
Box 800
Boulder, CO 80306
(800) 427-7680
Web site: www.springhillmedia.com

Songs for Sensory Integration
With Lois Hickman, M.S., O.T.R., and Aubrey Lande, M.S., O.T.R.
Belle Curve Records
P.O. Box 18387
Boulder, CO 80308
(800) 357-5067
An audiotape and booklet combining wonderfully fun children's music with developmentally enriched play activities for improving fine and gross motor skills, oral skills, and dressing and self-care. Booklet provides additional activity ideas and resources for parents dealing with SI, ADD/ADHD, LD, and related issues.

The Transitions™ Womb Sound Music Series
Transitions Music
P.O. Box 965
Snellville, GA 30078
(800) 492-9885
Web site: www.transitionmusic.com
Dr. Fred Schwartz's series on prenatal music.

PUBLICATION SOURCES

Don Campbell, Inc.
P.O. Box 4179
Boulder, CO 80306
(800) 721-2177
Web site: www.mozarteffect.com
For information on seminars, workshops, and classes by Don G. Campbell, as well as books and tapes by mail order.

Kindling Touch Publications
Dee Coulter, Director
4850 Niwot
Longmont, CO 80503
(303) 530-2357
Tapes and publications on the brain and music development for children.

MMB Music, Inc.
Contemporary Arts Building
3526 Washington Avenue
St. Louis, MO 63103
(314) 531-9635
Specialized books on music, health, and education.

Mozart Effect Resource Center
3526 Washington Avenue
St. Louis, MO 63103
(800) 721-2177
Web site: www.mozarteffect.com
Books and music education tools.

Zephyr Press, Inc.
3316 N. Chapel Avenue
P.O. Box 66006-MB
Tucson, AZ 85728-6006
(800) 232-2187
Fax: (520) 323-9402
Web site: www.zephyrpress.com
Books and curricula for multiple intelligences and brain body development.

PERIODICALS

Early Childhood Connections
P.O. Box 4274
Greensboro, NC 27404-4274
(910) 272-5303

Hearing Health
The Voice on Hearing Issues
P.O. Box 263
Corpus Christi, TX 78403
Bimonthly newsletter with many interesting articles on music and health.

WEB SITES AND DATABASES

ABC's of Parenting
www.abcparenting.com
Directory of parenting, pregnancy, and family Web sites, screened and selected by stay-at-home moms.

ArtsEdge
www.artsedge.kennedy-center.org/
Links arts and K-12 education through technology.

CAIRSS for Music
http://galaxy.einet.net/hytelnet/FUL064.html
A bibliographic database of music research literature. A joint venture between the faculty and staff of the University of Texas, San Antonio, and Southern Methodist University.

The Children's Group
www.childrensgroup.com
Recordings, research, and other valuable links on the benefits of classical music for children.

Education at the Met
www.operaed.org/
Tap the resources of the Metropolitan Opera.

Educational Resources Information Center
http://ericir.syr.edu/
A source of data on unpublished government-sponsored research and other resources.

E-mail Music Exchange
www.iglou.com/xchange/music/index.html
Lets music students correspond with musicians from around the world.

Family Education Network
www.familyeducation.com
Favorably rated for its offerings of educational information as well as fun and ease of use.

Federal Resources for Educational Excellence
www.ed.gov/free
From the U.S. Department of Education, hundreds of free, federally supported education resources and references for educators.

Goals 2000 Arts Education Partnership
http://aep-arts.org
Promotes the essential role of arts education in enabling all students to succeed in school, life, and work.

Harvard University Music Library
http://www.rism.harvard.edu/MusicLibrary/InternetResources.html
A starting place for searching public databases on music and music education.

Indiana University, School of Music
http://www.music.indiana.edu/music_resources/
A site maintained by the university for music and music education links.

Internet Music Resource Guide
www.teleport.com/~celinex/music.shtml
Comprehensive listings on nearly everything music-related on the Web.

MERB/CMI
http://www.ffa.ucalgary.ca/merb/index.html
A bibliographic database of more than 28,000 resources in music and music education from thirty-one Canadian and international journals and other sources covering the years 1956–1998. The journals are fully indexed by title, author, and subject.

Moms Online
http://www.momsonline.com
http://www.oxygen.com
A friendly site that includes columns, chats, and updated materials.

Mozart Effect Resource Center
www.mozarteffect.com
Books and music education tools.

Music Education Search System
http://www.music.utah.edu/MESS/
A source of general information and indexes of journals, maintained by the University of Utah.

MuSICA (Music and Science Information Computer Archive)
http://www.musica.uci.edu
E-mail: mbic@mila.ps.uci.edu
Thousands of regularly updated entries regarding music-related scientific research.

The New York Philharmonic Kidzone
www.nyphilkids.org
A child's-eye tour of Avery Fisher Hall, offering Q & A sessions with orches-

tra members, instructions on how to make musical instruments at home, and information about the conductors who have led the Philharmonic through the past century.

ParenthoodWeb
www.parenthoodweb.com
An award-winning site providing parenting-related articles, discussions, and an "Ask the Pros" section.

Parents.com
www.parents.com
Covers the common issues of parenting.

ParentsPlace.com
www.parentsplace.com
A "change-your-life," parent-helping-parent community.

Parent Soup
www.parentsoup.com
A nourishing blend of information for parents.

ParentZone
www.parentzone.com
A parenting resource that includes the *Momversations* newsletter.

Research Perspectives in Music Education
www.arts.usf.edu/music/rpme.html
A comprehensive collection of articles from research journals on music education.

NOTES

CHAPTER 1: TWINKLE TWINKLE, LITTLE NEURON

7 *Variations on Ah! Vous dirai-je, Maman* (K. 265): In the nineteenth century, each of Mozart's compositions was assigned a chronological number by the Austrian musicologist Ludwig von Köchel. Since then, the catalog of Mozart's work has been revised several times, but each composition still bears a Köchel listing.

9 sound in the early universe: R. Cowen, "Sound Waves May Drive Cosmic Structure," *Science News* 151, 11 January 1997, p. 21.

10 communication evolved from singing: Bruce Richman, "On the Evolution of Speech: Singing as the Middle Term," *Current Anthropology* (34), 1993, pp. 721–22. Cited in N. M. Weinberger, "Sing, Sing, Sing!" *MuSICA Research Notes* 3(2), Fall 1996, p. 2.

10 computer model of the brain: William F. Allman, "The Musical Brain," *U.S. News & World Report,* 11 June 1990, pp. 56–62.

12 "Exception among exceptions . . .": Alfred A. Tomatis, M.D., *Pourquoi Mozart?* (Paris: Editions Fixot, 1991).

14 New brain imaging technologies: Rima Shore, *Rethinking the Brain: New Insights into Early Development* (New York: Families and Work Institute, 1997), pp. 7–23.

15 Music can calm or stimulate: Michele Clements, "Observations on Certain Aspects of Neonatal Behavior in Response to Auditory Stimuli," paper presented at 5th International Congress of Psychosomatic Obstetrics and Gynecology, Rome, 1977. Cited in Thomas Verny, M.D., *The Secret Life of the Unborn Child* (New York: Dell, 1981), p. 39.

15 Premature infants who listen: Fred J. Schwartz, M.D., "Perinatal Stress Reduction, Music and Medical Cost Savings," *Journal of Prenatal and Perinatal Psychology and Health* 12(1), Fall 1997, pp. 19–29.

15 Young children who receive: M. Kalmar (1982), "The Effects of

Music Education Based on Kodály's Directives in Nursery School Children: From a Psychologist's Point of View," *Psychology of Music, Special Issue,* pp. 63–68. Also I. Hurwitz, P. H. Wolff, B. D. Bortnick, and K. Kokas (1975), "Nonmusical Effects of the Kodály Music Curriculum in Primary Grade Children," *Journal of Learning Disabilities* 8(3), 1975, pp. 167–74. Cited in *Music and Child Development,* J. Craig Peery and Irene Weiss Peery, eds. (Springer Verlag, 1987), p. 22. Also M. F. Gardiner, A. Fox, F. Knowles, and D. Jeffrey (1996), "Learning Improved by Arts Training," *Nature* 381(6), p. 284.

15 High school students who sing: "Profiles of SAT and Achievement Test Takers 1998," The College Board.

15 College students who listen: Frances H. Rauscher, Gordon L. Shaw, and Katherine N. Ky, "Listening to Mozart Enhances Spatial-Temporal Reasoning: Towards a Neurophysiological Basis," *Neuroscience Letters* (185), 1995, pp. 44–47.

15 Adult musicians' brains: Johnson, Petsche, Richter, von Stein, and Filz, *Music Perception* (13), 1996, pp. 563–82. Cited in "Brain Coherence, Musicianship and Gender," *MuSICA Research Notes* 3(2), Fall 1996, p. 8.

15 and even differ anatomically: G. Schlaug, L. Jancke, Y. Huang, J. F. Staiger, and H. Steinmetz, (1995), "Increased Corpus Callosum Size in Musicians," *Neuropsychologia* (33), pp. 1047–55. Also Sandra Blakeslee, "Piano Practice Alters Brain," *New York Times,* 8 May 1998.

CHAPTER 2: MOZART LISTENED TO MOZART

19 when his wife, Constanze, was pregnant: F. B. Lammes and M. M. Van, "The Labour of Constanze Mozart—a Musical Occasion?" *European Journal of Obstetrics, Gynecology, and Reproductive Biology* 37(1), 1990, pp. 41–46.

20 The human fetus is therefore capable of learning: Jean-Pierre Lecanuet, "Prenatal Auditory Experience," in Irène Deliège and John Sloboda, eds., *Musical Beginnings: Origins and Development of Musical Competence,* (New York: Oxford University Press, 1996), p. 24.

21 Beethoven's Fifth Symphony reaches the fetal ear: R. M. Abrams, K. Griffiths, and X. Huang, et al. (1998), "Fetal Music Perception: The Role of Sound Transmission," *Music Perception* (15), pp. 307–17.

Cited in N. M. Weinberger, "Lessons of the Music Womb," *MuSICA Research Notes* 6(1), Winter 1999, p. 2.

21 New sounds tend to cause the baby's heart rate: J.-P. Lecanuet, C. Granier-Deferre, and M.-C. Bunel (1988), "Fetal Cardiac and Motor Responses to Octave-Band Noises as a Function of Cerebral Frequency, Intensity and Heart Rate Variability," *Early Human Development* (18), pp. 81–93. Cited in N. M. Weinberger, "Lessons of the Music Womb," *MuSICA Research Notes* 6(1), Winter 1999, p. 2.

22 developing fetus behaves in just the same way: L. R. Leader, P. Baillie, and B. Martin, et al. (1982). "The Assessment and Significance of Habituation to a Repeated Stimulus by the Human Fetus," *Early Human Development* (7), pp. 211–19. Cited in N. M. Weinberger, "Lessons of the Music Womb," *MuSICA Research Notes* 6(1), Winter 1999, p. 2.

22 *associating* one event with another: David K. Spelt (1948). "The Conditioning of the Human Fetus in Utero," *Journal of Experimental Psychology* (38), pp. 338–47. Cited in N. M. Weinberger, "Lessons of the Music Womb," *MuSICA Research Notes* 6(1), Winter 1999, p. 3.

22 newborns clearly recognize and prefer music: B. J. Satt (1984), "An Investigation into the Acoustical Induction of Intrauterine Learning," *Dissertation Abstracts.* Unpublished doctoral dissertation, California School of Professional Psychology. Cited in N. M. Weinberger, "Lessons of the Music Womb," *MuSICA Research Notes* 6(1), Winter 1999, p. 4. Also Peter G. Hepper, "The Musical Foetus," *Irish Journal of Psychology* (12), 1991, pp. 95–107.

22 Babies can even develop *literary* preferences: Gina Kolata, "Rhymes Reason: Linking Thinking to Train the Brain?" *New York Times,* 19 February 1995.

22 newborns can tell the difference between: Jean-Pierre Lecanuet, "Prenatal Auditory Experience," in Irène Deliège and John Sloboda, eds., *Musical Beginnings: Origins and Development of Musical Competence* (New York: Oxford University Press, 1996), p. 26.

22 music can be used to enhance their development: M. J. LaFuente, R. Grifol, and J. Segarra, et al. (1997), "Effects of the Firstart Method of Prenatal Stimulation on Psychomotor Development: The First Six Months," *Prenatal and Perinatal Psychology Journal* 11(3), pp. 151–62.

24 During World War II: Grace C. Nash, "Thoughts About Musical Development," *Early Childhood Connections,* Winter 1995, pp. 35–38.

24 Her general level of stress: T. Field, "Music Interventions May Reduce Maternal Depression and Its Effects on Infants," *Preventive Medicine* (27), 1998, pp. 200–03.

25 frequent or long-lasting emotional states: Matti O. Huttunen and Niskanen Pekka, "Prenatal Loss of Father and Psychiatric Disorders," *Archives of General Psychiatry*, April 1978, pp. 429–31.

25 more underweight and frequently crying babies: A. J. Crandon (1979a), "Maternal Anxiety and Neonatal Well-Being," *Journal of Psychosomatic Research* (23), pp. 113–15.

25 normal, minor stresses of pregnancy: Nicolino Rossi, Paola Avveduti, Nicola Rizzo, and Raffaele Lorusso, "Maternal Stress and Fetal Motor Behavior: A Preliminary Report," *Prenatal and Perinatal Psychology Journal* 3(4), Summer 1989, pp. 311–17.

26 Recent research in cell biology: Bruce H. Lipton, Ph.D., "Nature, Nurture and the Power of Love," *Prenatal and Perinatal Psychology Journal* 13(1), Fall 1998, pp. 3–10.

27 "For a long time other scientists believed . . .": Interview with Dr. Thomas Verny, February 1999.

28 In Valencia, Spain: Rosario N. Rozada Montemurro, "Singing Lullabies to Unborn Children: Experiences in Village Vilamarxant, Spain," *Prenatal and Perinatal Psychology Journal* 11(1), Fall 1996, pp. 9–16.

32 one or two sessions of low-volume classical violin music: Donald J. Shetler, Ed.D., "The Inquiry into Prenatal Musical Experience: A Report of the Eastman Project 1980–1987," *Prenatal and Perinatal Psychology Journal* 3(3), Spring 1989, pp. 171–89. Also M. J. Lafuente, R. Grifol, J. Segarra, J. Soriano, et al., "Effects of the Firstart Method of Prenatal Stimulation on Psychomotor Development: The First Six Months," *Prenatal and Perinatal Psychology Journal* 11(3), pp. 151–62.

33 "prenatal university": F. Rene Van de Carr, M.D., and Marc Lehrer, Ph.D., "Prenatal University: Commitment to Fetal-Family Bonding and the Strengthening of the Family Unit as an Educational Institution," *Prenatal and Perinatal Psychology Journal* 3(2), Winter 1988, pp. 87–102.

35 three-tone melody: Grace C. Nash, "Thoughts About Musical Development," *Early Childhood Connections*, Winter 1995, pp. 35–38.

36 Some of Dr. Min's patients: The Associated Press, "Apron Is Music to the Ears of the Unborn," *The New Mexican*, 2 February 1998, p. A-6.

37 prenatal exposure to music: Donald J. Shetler, Ed.D., "The Inquiry into Prenatal Musical Experience: A Report of the Eastman Project 1980–1987," *Prenatal and Perinatal Psychology Journal* 3(3), Spring 1989, pp. 171–89.

38 pregnant women who smoked: Michael Lieberman, "Gravida's Smoking Seen as Handicap to Offspring," *Obstetrics-Gynecology News* 5(12), 15 June 1970, p. 16.

39 one community in Uganda: Jack Kornfield, *A Path with Heart: A Guide through the Perils and Promises of Spiritual Life* (New York: Bantam, 1996), p. 120.

41 Recordings with distinct, even rhythms: Janie C. Livingston, RN, Med, "Principles and Practice: Music for the Childbearing Family," *JOGN Nursing,* November/December 1979, pp. 363–67.

42 music while undergoing labor: Carlos E. Gonzales, "The Music Therapy–Assisted Childbirth Program: A Study Evaluation," *Prenatal and Perinatal Psychology Journal* 4(2), Winter 1989, pp. 111–24.

43 Dr. Fred Schwartz: Fred J. Schwartz, M.D., "Using Music for Relaxation During Labor," *IMSPD Newsletter* 9(1), Fall 1996. Cited in Loran Zemke, OSF, DMA, "Music and Prenatal Development," *Early Childhood Connections,* Summer 1998, pp. 19–23.

43 half as likely to need traditional anesthesia: Marian Westley, "Music Is Good Medicine," *Newsweek,* 21 September 1998, p. 102.

43 helps women focus and cope with pain: Beverly Pierce, "The Practice of Toning in Pregnancy and Labour: Participant Experiences," *Complementary Therapies in Nursing and Midwifery* (4), 1998, pp. 41–46.

44 as a woman moves more deeply into labor: M. Odent, *Birth Reborn* (New York: Pantheon, 1994).

CHAPTER 3: CRY BABY, LULLABY, AND ITSY-BITSY SONGS

46 By two months of age: Dr. Jayne Standley and Clifford K. Madsen, "Comparison of Infant Preferences and Responses to Auditory Stimuli," *Journal of Music Therapy* 27(2), 1990, pp. 54–97.

46 By five months: L. W. Olsho, "Infant Frequency Discrimination," *Infant Behavior and Development* (7), 1984, pp. 27–35.

46 at six months: B. Bower, "Infants Tune Up to Music's Core Qualities," *Science Notes,* 7 September 1996, p. 151.

46 recognize familiar melodies even when the tempo is changed: S. E.

Trehub and L. A. Thorpe, "Infants' Perception of Rhythm: Categorization of Auditory Sequences by Temporal Structure," *Canadian Journal of Psychology* (43), 1989, pp. 217–29.

46 to recognize a wrong note: William F. Allman, "The Musical Brain," *U.S. News & World Report,* 11 June 1990, pp. 615–26.

47 understand music according to the rules: L. J. Trainor and S. E. Trehub, "Musical Context Effects in Infants and Adults: Key Distance," *Journal of Experimental Psychology: Human Perception and Performance* (19), 1993, pp. 615–26.

47 during the first few weeks after birth: B. Rosner and A. Doherty, "The Response of Neonates to Intrauterine Sounds," *Developmental Medicine and Child Neurology* (21), 1979, pp. 723–29. Also, Sharon K. Collins, RNC, MN, and Kay Kuck, RN, MN, "Music Therapy in the Neonatal Intensive Care Unit," *Neonatal Network* 9(6), March 1991, pp. 23–26.

47 the weight of "preemies" increased faster: "Healing Unlimited," *The Healthy Family,* Fall 1994, p. 181.

47 reduces the hospital stay of premature and low-birth-weight babies: Fred J. Schwartz, M.D., "Perinatal Stress Reduction, Music and Medical Cost Savings," *Prenatal and Perinatal Psychology Journal* 12(1), Fall 1997, pp. 19–29.

48 calming and stabilizing effects of music: Colleen A. Lorch, Vichien Lorch, Allan O. Diefendorf, and Patricia W. Earl, "Effect of Stimulative and Sedative Music on Systolic Blood Pressure, Heart Rate, and Respiratory Rate in Premature Infants," *Journal of Music Therapy* (31), 1994, pp. 105–118.

49 the sounds of words build up neural circuitry: Edwin E. Gordon, "Early Childhood Music Education: Life or Death? No, a Matter of Birth and Life," *Early Childhood Connections,* Fall 1996, pp. 7–11.

52 an auditory map of your language: Sharon Begley, "How to Build a Baby's Brain," *Newsweek,* 19 February 1996, pp. 28–31.

52 the more emotional the exchange: Ibid.

53 Holding and cuddling a baby: Sarah Van Boven, "Giving Infants a Helping Hand," *Your Child from Birth to Three, Newsweek Special Issue,* Spring/Summer 1997, p. 45.

53 Severe damage to an infant's inner ear: J. Greenberg, "Early Hearing Loss and Brain Development," *Science News* 131(10), 7 March 1987, p. 149.

53 consider the effects of the twenty-four-hour-per-day sounds: Jane W. Cassidy, Ph.D., and Karen M. Ditty, M.S., CCC-A, "Presentation of Aural Stimuli to Newborns and Premature Infants: An Audiological Perspective," *Journal of Music Therapy,* 35(2), 1998, pp. 70–87.

55 parents sing their babies through playtime: Sandra E. Trehub, "The World of Infants: A World of Music," *Early Childhood Connections,* Fall 1996, pp. 27–34.

57 A strong, secure attachment to a nurturing caregiver: Rima Shore, *Rethinking the Brain: New Insights into Early Development* (New York: Families and Work Institute, 1997), pp. 27–29.

58 musically untrained children: C. Robazza, C. Macaluso, and V. D'Urso, (1994), "Emotional Reactions to Music by Gender, Age, and Expertise." *Perceptual and Motor Skills* (79), pp. 939–44. Cited in N. M. Weinberger, "Elevator Music: More Than It Seems," *MuSICA Research Notes* 2(2), Fall 1995, p. 5.

59 the process of attunement within the family: Daniel N. Stern, M.D., *Diary of a Baby: What Your Child Sees, Feels, and Experiences* (New York: Basic Books, 1998). Cited in Sharon Begley, "Your Child's Brain," *Newsweek,* 19 February 1996, pp. 55–62.

61 baby's response when an adult sings: Dr. Beth M. Bolton, "Was That a Musical Response? Eliciting and Evaluating Musical Behaviors in Very Young Children," *Early Childhood Connections,* Fall 1996, pp. 14–18.

62 Cognitive scientist Peter Juscyk: Geoffrey Cowley, "The Language Explosion," *Your Child from Birth to Three, Newsweek Special Issue,* Spring/Summer 1997, pp. 16–21.

63 infants can distinguish rhyming verse: Gina Kolata, "Rhyme's Reason: Linking Thinking to Train the Brain?" *New York Times,* 19 February 1995, p. E3.

63 adults singing to an infant: L. J. Trainor, (1996), "Infant Preferences for Infant-Directed Versus Non-Infant-Directed Play Songs and Lullabies," *Infant Behavior and Development* (19), pp. 83–92.

64 biological process called vocal marking: Hanus Papoušek, "Musicality in Infancy Research: Biological and Cultural Origins of Early Musicality," *Musical Beginnings,* Irène Deliège and John Sloboda, eds. (New York: Oxford University Press, 1996), pp. 37–51.

64 musical aspects of baby talk: Ibid.

64 When talking to your baby: Geoffrey Cowley, "The Language Explo-

sion," *Your Child from Birth to Three, Newsweek Special Issue,* Spring/ Summer 1997, pp. 16–23.

68 infants whose mothers speak to them a lot: Sharon Begley, "How to Build a Baby's Brain," *Your Child from Birth to Three, Newsweek Special Issue,* Spring/Summer 1997, p. 31. Citing Janellen Huttenlocher of the University of Chicago.

68 If musical ability is also a high priority: Edwin E. Gordon, *A Music Learning Theory for Newborn and Young Children* (Chicago: Gia Publications, Inc., 1997).

69 Heartbeat and breathing rates, motor movements, and brain waves: Jeff Strong, "Rhythmic Entrainment Intervention as a Treatment Option for Individuals with Autism," press release, REI Institute, Inc., www.reiinstitute.com.

CHAPTER 4: CRAWL, REACH, AND CLAP

75 About the time she begins sitting up: Louie Suthers, "Infant and Toddler Movement Responses to Sound-Making," *Early Childhood Connections,* Winter 1997, pp. 19–20.

77 the speed and efficiency with which your baby accomplishes such movements: Karen E. Adolph, Beatrix Vereijken, and Mark A. Denny, "Learning to Crawl," *Child Development* 69(5), October 1998, pp. 1299–1312.

78 Once mind and body begin to meld: Dee J. Coulter, *Beginning with Music* (audiotape).

79 language's sounds and rhythms: Gina Kolata, "Rhyme's Reason: Linking Thinking to Train the Brain?" *New York Times,* 19 February 1995, p. E3.

80 34 to 53 percent of the infants' vocal sounds: Mechthild Papoušek, "Intuitive Parenting: A Hidden Source of Musical Stimulation in Infancy," *Musical Beginnings,* Irène Deliège and John Sloboda, eds. (New York: Oxford University Press, 1996), pp. 88–109.

80 baby's musical development progresses right alongside her language growth: P. Ostwald, "Musical Behavior in Early Childhood," *Developmental Medicine and Child Neurology* (15), 1973, pp. 375–76.

81 Your own singing, with a full spectrum: Mechthild Papoušek, "Intuitive Parenting: A Hidden Source of Musical Stimulation in Infancy,"

Musical Beginnings, Irène Deliège and John Sloboda, eds. (New York: Oxford University Press, 1996), pp. 88–109.

81 infants just three months old best remembered how to manipulate a crib mobile: Fagan, Prigot, Carroll, Pioli, Stein, and Franco, *Child Development* (68), 1997, pp. 1057–66.

82 "Children's songs are an excellent illustration . . .": Paul Madaule, "Music: An Invitation to Listening, Language and Learning," *Early Childhood Connections,* Spring 1997, pp. 30–34.

82 By making sure to speak and sing with a wide range: S. J. Lamb, and A. H. Gregory, "The Relationship Between Music and Reading in Beginning Readers," *Educational Psychology* (13), pp. 19–26.

82 Read picture songbooks: J. Craig Peery, and Irene Weiss Peery, "The Role of Music in Child Development." Cited in J. C. Peery, et al., eds., *Music and Child Development* (New York: Springer-Verlag, 1987), pp. 3–31.

83 Movement grows the brain: Phyllis Weikart, "Purposeful Movement: Have We Overlooked the Base?" *Early Childhood Connections,* Fall 1995, pp. 6–15.

84 one hundred Michigan teenagers: Ibid. For further information, see P. S. Weikart, L. J. Schweinhart, and M. Larner, "Movement Curriculum Improves Children's Rhythmic Competence," *High/Scope Resource* 6(1), Winter 1987, pp. 8–10.

85 link between physical action and language: Rae Pica, "Moving and Learning Across the Curriculum," *Early Childhood Connections,* Spring 1996, pp. 22–29.

85 television does indeed reinforce listening and movement skills: Katharine Smithrim, "Vital Connections: Young Children, Adults, and Music." Paper presented at the ISME Early Childhood Commission Seminar, 11–15 July 1994, University of Missouri-Columbia. Cited in Joyce Jordan-DeCarbo, "Research Reviews," *Early Childhood Connections,* Winter 1996, p. 39.

86 only live language, not television: Sharon Begley, "How to Build a Baby's Brain," *Newsweek,* 19 February 1996, pp. 28–31.

87 "You're increasing by 50 percent . . .": Dee J. Coulter, *Meeting the World with Music* (audiotape).

89 group musical activities that include singing: C. Hoskins, "Use of Music to Increase Verbal Response and Improve Expressive Language Abilities of Preschool Language-Delayed Children," *Journal of Music*

Therapy (25), 1988, pp. 73–84. Also M. Kalmar, "The Effects of Music Education Based on Kodály's Directives in Nursery School Children: From a Psychologist's Point of View," *Psychology of Music, Special Issue,* 1982, pp. 63–68.

89 "It is as if there were a little secret that is only ours . . .": Josette Silveira Mello Feres, "Music as a Mediating Element in the Mother-Baby Relationship," *Early Childhood Connections,* Winter 1996, pp. 21–25.

90 "True success is achieved when the parents . . .": Kenneth K. Guilmartin, "Parent Inclusion: Nurturing the Musical Family," press release, Center for Music and Young Children, 1998, pp. 1–15.

92 three out of every one hundred babies: Robin Riccitiello and Jerry Alder, "Your Baby Has a Problem," *Your Child from Birth to Three, Newsweek Special Issue,* Spring/Summer 1997, pp. 46–50.

92 You may have noticed: Geoffrey Cowley, "The Language Explosion," *Your Child from Birth to Three, Newsweek Special Issue,* Spring/Summer 1997, pp. 16–22.

92 "Gabe is three years old . . .": Katherine C. Morehouse, M.Ed., CCC-A, "So Your Child Can't Hear . . . Consider the Possibilities," *Hearing Health,* July/August 1998, pp. 12–38.

93 improvisational music therapy improves autistic children's communicative behaviors: C. L. Edgerton, "The Effect of Improvisational Music Therapy on the Communicative Behaviors of Autistic Children," *Journal of Music Therapy,* 1994, pp. 31–62. Cited in *MuSICA Research Notes* 1(2), Fall 1994, p. 10.

93 Nordoff-Robbins Music Therapy Clinics: Brochure, The Nordoff-Robbins Music Therapy Foundation, Inc.

94 Rhythmic entrainment intervention (REI) is also used frequently: Jeff Strong, "Rhythmic Entrainment Intervention as a Treatment Option for Individuals with Autism," Vista, Calif.: REI Institute, Inc.

CHAPTER 5: DANCE AND PLAY

97 Rhythm, pitch, lyrics, and tone will structure his auditory experience: Donald E. Michel, Ph.D., RMT-BC, and Janet L. Jones, M.A., RMT-BC, *Music for Developing Speech and Language Skills in Children: A Guide for Parents and Therapists* (St. Louis: MMB Music, Inc., 1991), pp. 7–27.

98 words are not everything: Daniel N. Stern, M.D., *Diary of a Baby: What Your Child Sees, Feels, and Experiences* (New York: Basic Books, 1990), p. 114.

99 Unused neuronal connections die out in droves: Rima Shore, *Rethinking the Brain: New Insights into Early Development* (New York: Families and Work Institute, 1997), pp. 18–20.

100 5 percent of toddlers: Dee J. Coulter, *A Guided Tour of the Brain* (audiotape).

100 "My musician friends . . .": Dee J. Coulter, "Defending the Magic: Current Issues in Early Childhood Education," *Early Childhood Connections,* Spring 1996, pp. 30–38.

100 they play games that deal with the structuring of sounds: L. B. Miller, "Children's Musical Behaviors in the Natural Environment." Cited in J. C. Peery, et al., eds., *Music and Child Development* (New York: Springer-Verlag, 1987), pp. 206–24.

102 researcher Karen Wolff: K. Wolff, (1979), "The Effects of General Music Education on the Academic Achievement, Perceptual-Motor Development, Creative Thinking, and School Attendance of First-Grade Children," unpublished doctoral dissertation, University of Michigan, 1979. *Dissertation Abstracts International,* 40, 5359A. Cited in Robert Cutietta, Donald L. Hamann, and Linda Miller Walker, eds., *Spin-Offs: The Extra-Musical Advantages of a Musical Education* (Ekhart, Ind.: United Musical Instruments U.S.A., Inc., for the Future of Music Project, 1995).

103 When you play a piece of recorded music: Phyllis Weikart, "Purposeful Movement: Have We Overlooked the Base?" *Early Childhood Connections,* Fall 1995, pp. 6–15.

103 moving to music gives kids a chance to develop basic timing: Elizabeth H. Carlton, "Building the Musical Foundation: Music Key Experiences in Active Learning Settings," *Early Childhood Connections,* Fall 1995, pp. 16–21.

104 active involvement with music is more effective: Elaine Metz, "Movement as a Musical Response among Preschool Children," *Journal of Research in Music Education* (37), 1989, pp. 48–60.

105 time of your child's second birthday: Donald E. Michel, Ph.D., RMT-BC, and Janet L. Jones, M.A., RMT-BC, *Music for Developing Speech and Language Skills in Children: A Guide for Parents and Therapists* (St. Louis: MMB Music, Inc., 1991), pp. 7–27.

106 For true communication to take place: Conversation with Dr. Dee J. Coulter, May 1999.

107 you can clap out the beat: Read*Write*Now, Materials for Kids from Birth to Grade 1, www.ed.gov.

107 the "exquisite dance" that good conversationalists engage in: Deborah Flores, "Foundations of Learning," *Parenting,* www.family.com

110 Spurred on by the sounds: Elizabeth Carlton, "Building the Musical Foundation: Music Key Experiences in Active Learing Settings," *Early Childhood Connections,* Fall 1995, pp. 16–21.

110 Six percent of fourth graders: D. K. Lipscomb, *Hearing Conservation in Industry, Schools and the Military* (San Diego, Calif.: Singular Publishing Group, Inc., 1994). Cited in Nancy Nadler, M.E.D., M.A., "Noisy Toys: Hidden Hazards," *Hearing Health,* November/December 1997, pp. 18–21.

112 between ten and eighteen months of age: Sharon Begley, "Your Child's Brain," *Newsweek,* 19 February 1996, pp. 55–62.

113 your child has been testing his own importance: T. Berry Brazelton, "Building a Better Self-Image," *Your Child from Birth to Three, Newsweek Special Issue,* Spring/Summer 1997, pp. 76–77.

113 children's natural aversion to dentistry: Robert A. Wortzel, D.M.D., *A Trip to the Dentist Can Be Lots of Fun!* (videotape).

115 Experimenting with what it's like to be someone or something else: Debra Rosenberg, "Raising a Moral Child," *Your Child from Birth to Three, Newsweek Special Issue,* Spring/Summer 1997, pp. 92–93.

116 structured group music instruction increases vocabulary: C. Hoskins, "Use of Music to Increase Verbal Response and Improve Expressive Language Abilities of Preschool Language-Delayed Children," *Journal of Music Therapy* (25), 1988, pp. 73–84.

116 motor development, particularly coordination, abstract conceptual thinking, play improvisation, originality: M. Kalmar, "The Effects of Music Education Based on Kodály's Directives in Nursery School Children: From a Psychologist's Point of View," *Psychology of Music, Special Issue,* 1982, pp. 63–68.

116 and social interaction in young children: Joyce Jordan-DeCarbo, "The Role of Music in Enhancing Social Interactions among Preschoolers," *Early Childhood Connections,* Spring 1998, p. 43.

118 music reduces the rate of inappropriate or disruptive behavior: Claire V. Wilson and Leona S. Aiken, "The Effect of Intensity Levels upon

Physiological and Subjective Affective Response to Rock Music," *Journal of Music Therapy* (14), 1977, pp. 60–77. Also, Robert Cutietta, Donald L. Hamann, and Linda Miller Walker, eds., *Spin-Offs: The Extra-Musical Advantages of a Musical Education* (Ekhart, Ind.: United Musical Instruments U.S.A., Inc., for the Future of Music Project, 1995), pp. 29–32.

118 music has a powerful effect not only on mood: T. Taniguchi, "Mood Congruent Effect by Music on Word Recognition" (Japanese lang.), *Shinrigaku Kenkyu* (62), 1991, pp. 88–95; G. Chastain, P. S. Seilbert, and F. R. Ferraro, "Mood and Lexical Access of Positive, Negative, and Neutral Words," *Journal of General Psychology* (122), 1995, pp. 137–57; V. N. Stratton, and A. H. Zalanowski, "The Effects of Music and Paintings on Mood," *Journal of Music Therapy* (26), 1989, pp. 30–41. All cited in N. M. Weinberger, "The Coloring of Life: Music and Mood," *MuSICA Research Notes* 3(1), Spring 1996, pp. 4–6.

122 Group music activity has been shown: L. Miller, "A Description of Children's Musical Behaviors: Naturalistic," *Bulletin of the Council for Research in Music Education* (87), 1986, pp. 1–16.

123 "the attachment between grandparent and grandchild . . .": Kenneth L. Woodward, "A Grandparent's Role," *Your Child from Birth to Three, Newsweek Special Issue,* Spring/Summer 1997, pp. 81–83.

CHAPTER 6: TAP, RAP, AND SING ALONG

128 As children begin to express themselves: Gladys Evelyn Moorhead and Donald Pond, "Music of Young Children," *Early Childhood Connections,* Summer 1997, pp. 43–47.

129 By asking such questions now as: Adapted from "Read*Write*Now," Materials for Kids from Birth to Grade 1, www.ed.gov.

130 naturally and freely experiment: L. B. Miller, "Children's Musical Behaviors in the Natural Environment," *Music and Child Development,* J. C. Peery, et al., eds. (New York: Springer-Verlag, 1989), pp. 206–24. Cited in N. M. Weinberger, "Sing, Sing, Sing!" *MuSICA Research Notes* 3(2), Fall 1996, pp. 1–4.

130 very clear organizational patterns: H. A. Veldhuis, "Spontaneous Songs of Preschool Children," *Arts in Psychotherapy* (11), (1984), pp. 15–24.

132 Gradually, she will find that her songs: Jane M. Healy, "Children's Brains at Work in the Preschool and Elementary Years," *Early Childhood Connections,* Spring 1996, pp. 7–16.

133 Sequence songs such as "The Hokey Pokey": B. Rowen, *Learning Through Movement* (New York: Teachers College, 1982).

135 Join her in music making: Viviane Pouthas, "The Development of the Perception of Time and Temporal Regulation of Action in Infants and Children," *Musical Beginnings,* Irène Deliège and John Sloboda, eds. (New York: Oxford University Press, 1996), pp. 131–39.

135 When young Mozart was four years old: *Mozart* (CD-ROM) liner notes, The Multimedia Composer Series.

136 greater creativity: N. M. Weinberger, "Creating Creativity with Music," *MuSICA Research Notes* 5(2), Spring 1998, pp. 1–5.

136 a group of economically disadvantaged children: "Music in Preschool Speeds Tots' Mental Progress," *Art and Music Brain/Mind Collections* 18(11C), 1995. Findings were reported at the annual meeting of the National Association of Music Merchants in Newport Beach, Calif. Reported on CBC-TV's *The National,* 17 February 1998.

136 first-graders who were given thirty minutes of daily music instruction: K. J. Wolff, "The Effects of General Music Education on the Academic Achievement, Perceptual-Motor Development, Creative Thinking, and School Attendance of First-Grade Children," doctoral dissertation, University of Michigan, 1979. *Dissertation Abstracts International,* 40, 5359A. Cited in N. M. Weinberger, "Creating Creativity with Music," *MuSICA Research Notes* 5(2), Spring 1998, pp. 1–5.

136 more coherence between the hemispheres of the brain: Johnson, Petsche, Richter, von Stein, and Filz, *Music Perception* (13), 1996, pp. 563–82.

136 have a better memory for words: Agnes Chang, et al., "Music Training Improves Verbal Memory," Chinese University, Hong Kong. Reported in *New Scientist,* 4 July 1999.

136 actively making music has a substantially greater beneficial effect: Francis H. Rauscher, Ph.D., "What Educators Must Learn from Science: The Case for Music in the Schools," *Early Childhood Connections,* Spring 1996, pp. 17–21.

138 more than 300,000 students and teachers: Winifred Crock, "The Suzuki Philosophy," *Parent Education* (pamphlet, n.d.), Parkway School District Orchestras.

138 parental involvement in the training: "Study Affirms Early Music Training Improves Intelligence," *Sam Houston State University Media,* 7 June 1999.

139 help get her back on track: Cheryl Cornell, "Helping the Child with

Low Tolerance for Frustration," *American Suzuki Journal,* Summer 1997, pp. 97–98.

139 "Isn't it a thrill . . .": Elizabeth Mills, *The Suzuki Concept: An Introduction to a Successful Method for Early Music Education* (Berkeley, Calif.: Diablo Press, 1973), pp. 18–19.

139 "One of the most exciting things . . .": Ibid.

CHAPTER 7: SING . . . SING A SONG

141 When Mozart was only five: "Mozart's Young Life," tracks 6–19, *Mozart* (CD-ROM) liner notes, The Multimedia Composer Series.

142 a small part of the brain's parietal lobe: Jane M. Healy, Ph.D., "Children's Brains at Work in the Preschool and Elementary Years," *Early Childhood Connections,* Spring 1996, pp. 7–16.

143 Your child's brain is designed to help him adapt: David F. Bjorklund, "In Search of a Metatheory for Cognitive Development (or, Piaget Is Dead and I Don't Feel So Good Myself)," *Child Development* 68(1), February 1997, pp. 144–48.

143 three-year-olds who simply attended twice-weekly singing lessons: M. Kalmar, "The Effects of Music Education Based on Kodály's Directives in Nursery School Children: From a Psychologist's Point of View," *Psychology of Music, Special Issue,* 1982, pp. 63–68.

143 "Music is a primary language of the brain . . .": Peter Perret, "Sowing the Seeds: The Bolton Project," *Symphony,* January/February 1999, p. 42.

144 the concepts underlying geometry: Dee J. Coulter, "The Brain's Timetable for Developing Musical Skills," *Orff Echo* 14(3), Spring 1982, pp. 18–22.

144 the typical adult professional spends nearly 55 percent of his or her time: Don G. Campbell, and Chris Brewer, *Rhythms of Learning* (Tucson, Ariz.: Zephyr Press, 1991).

145 sixty million Americans suffer from this disability: Don Campbell, *The Mozart Effect* (New York: Avon Books, 1997).

147 The Poetry of Sound: For more information on how to use music to stimulate learning and the imagination, see Don Campbell, and Chris Brewer, *Rhythms of Learning* (Tucson, Ariz.: Zephyr Press, 1991).

149 playing folk or pop background music: M. R. Godeli, P. R. Santana, V. H. Souza, and G. P. Marquetti, "Influence of Background Music

on Preschoolers' Behavior: A Naturalistic Approach," *Perceptual and Motor Skills* (82), 1996, pp. 1123–29.

149 teaching technique called Improvised Musical Play: A. S. Gunsberg, "Play as Improvisation: The Benefits of Music for Developmentally Delayed Young Children's Social Play," *Early Child Development and Care* (66), 1991, pp. 85–91.

153 greater intuitive powers and self-confidence: Dee J. Coulter, *A Guided Tour of the Brain*, audiotape (Longmont, Colo.: Kindling Torch Publications, 1988).

153 Mind-body games such as this one: For more information on mind-body interaction and education, see Don G. Campbell, *Introduction to the Musical Brain,* 2nd ed. (St. Louis, Mo.: MMB Music Inc., 1983).

155 The neurological, socioemotional, and perceptual-motor benefits: Robert Cutietta, Donald L. Hamann, and Linda Miller Walker, eds., *Spin-Offs: The Extra-Musical Advantages of a Musical Education* (Ekhart, Ind.: United Musical Instruments U.S.A., Inc., for the Future of Music Project, 1995), pp. 29–32. Also, I. Hurwitz, P. H. Wolff, B. D. Bortnick, and K. Kokas, "Nonmusical Effects of the Kodály Music Curriculum in Primary Grade Children," *Journal of Learning Disabilities* 8(3), 1975, pp. 167–74. Cited in *Music and Child Development,* J. Craig Peery and Irene Weiss Peery, eds. (New York: Springer Verlag, 1987), p. 22.

155 the role of the kindergarten music teacher: Katalin L. Nagy, "Early Childhood Music Education in Hungary," *Early Childhood Connections,* Winter 1997, pp. 6–13.

156 "The songs and rhymes . . .": John Feierabend, Ph.D., "Music and Movement for Infants and Toddlers: Naturally Wonder-Full," *Early Childhood Connections,* Fall 1996, pp. 19–26.

157 relationship between body consciousness and intellectual activity: Rae Pica, "Moving and Learning across the Curriculum," *Early Childhood Connections,* Spring 1996, pp. 22–29.

158 To remedy this difficulty, she recommends "cross crawling": Carla Hannaford, *Smart Moves: Why Learning Is Not All in Your Head* (Arlington, Va.: Great Ocean Publishers, 1995).

159 attempted to teach the names of body parts: B. Mohanty and A. Hejmadi, "Effects of Intervention Training on Some Cognitive Abilities of Preschool Children," *Psychological Studies* (37), pp. 31–37, cited in N. M. Weinberger, "Music and Cognitive Achievement in Children," *MuSICA Research Notes* 1(2), Fall 1994, pp. 1–4.

159 Reading readiness involves a number of distinct skills: U. Frith, (1985), "Beneath the Surface of Developmental Dyslexia," in K. E. Patterson, J. C. Marshall, and M. Coltheart, eds. *Surface Dyslexia* (Hove, Lawrence Erlbaum Associates Ltd.), pp. 301–30. *Ibid.*

159 children who do well at telling: S. J. Lamb and A. H. Gregory (1993), "The Relationship Between Music and Reading in Beginning Readers," *Educational Psychology* (13), pp. 19–26. *Ibid.*

159 Music can also affect reading accuracy: C. M. Colwell, "Therapeutic Application of Music in the Whole Language Kindergarten," *Journal of Music Therapy* (31), 1994, pp. 238–47.

160 best to start big: Rae Pica, "Moving and Learning across the Curriculum," *Early Childhood Connections,* Spring 1996, pp. 22–29.

162 nearly a hundred remedial first graders: M. F. Gardiner, A. Fox, F. Knowles, and D. Jeffrey (1996), "Learning Improved by Arts Training," *Nature* 381(6), 580:284. Cited in N. M. Weinberger, *MuSICA Research Notes* 4(1), Spring 1997, p. 8.

162 increase in alpha rhythm frequency: N. M. Weinberger, "Music Alters Children's Brain Waves," *MuSICA Research Notes* 5(1), Winter 1998, p. 7.

163 music your child enjoys can create positive *physiological* effects: Erika Leeuwenburgh, MPS, ATR-BC, CCLS, and Kate Richards, M.A., MT-BC, "The Use of Music Listening in the Pediatric Intensive Care Environment: A Descriptive Study of the Physiological and Psychological Effects of Listening to Sedative Music," Hackensack University Medical Center Dept. of Pediatrics press release, 27 January 1999.

165 music has been proven to reliably alter young children's mood: "Survey, Music for Kids Suggest Methods of Mood Improvement," *New Sense Bulletin,* April 1992, pp. 2–3.

165 the students at Bohemia Elementary School: Associated Press, "Pupils Fund Class Themselves," *The Bulletin,* 23 March 1999.

CHAPTER 8: RHYTHMS OF THOUGHT

169 Even Peter Perret, music director: Peter Perret, "Sowing the Seeds: The Bolton Project," *Symphony,* January/February 1999, pp. 40–42.

172 Mozart loved math: *Mozart* (CD-ROM) liner notes, The Multimedia Composer Series.

173 first graders who participated in Kodály: I. Hurwitz, P. H. Wolff,

B. D. Bortnick, and K. Kokas (1975), "Nonmusical Effects of the Kodály Music Curriculum in Primary Grade Children," *Journal of Learning Disabilities* (8), pp. 45–51.

173 The ancient Greeks were highly familiar: Uschi Felix, "The Contribution of Background Music to the Enhancement of Learning in Suggestopedia: A Critical Review of the Literature," *Journal of the Society for Accelerative Learning and Teaching* 18(3,4), Fall & Winter 1993, pp. 277–303.

173 Renowned music researcher Jay Dowling: W. J. Dowling (1993), "Procedural and Declarative Knowledge in Music Cognition and Education," cited in T. J. Tighe and W. J. Dowling, eds., *Psychology and Music: The Understanding of Melody and Rhythm* (Hillsdale, N.J.: Lawrence Erlbaum Associates, 1992), pp. 5–18. Cited in N. M. Weinberger, "Music and Its Memories," *MuSICA Research Notes* 3(2), Fall 1996.

173 "singing stimulates the nerves to the vestibular system . . .": Conversation with Carla Hannaford, May 1999.

174 "The name of our president . . .": Robert Greene, "Rote Learning Method Draws Debate, Praise," *Boulder Camera,* 31 May 1998, p. 5A.

175 The Unnervous Tic: For more information on how to use music to aid learning at home and in the classroom, see Don G. Campbell, and Chris Brewer, *Rhythms of Learning* (Tucson, Ariz.: Zephyr Press, 1991).

176 In his fascinating book *When Listening Comes Alive*: Paul Madaule, *When Listening Comes Alive* (Norval, Ontario: Moulin Publishing, 1993).

177 even if your child's classroom seems quiet: Carl C. Crandell, Ph.D., "Classroom Acoustics: A Failing Grade," *Hearing Health,* Sept./Oct. 1998, pp. 11–59.

178 This aural stretching can especially improve: Paul Madaule, "Listening Training and Music Education," *Early Childhood Connections,* Spring 1998, pp. 35–41.

183 research supports this student's contention: Carly L. Price, "Is That All There Is?: A Number of Alternative Methods to Ritalin Offer Hope without Drugs," *Common Boundary,* March/April 1998, pp. 33–41.

183 nineteen children aged seven to seventeen: Don Campbell, *The Mozart Effect* (New York: Avon Books, 1997), p. 233.

184 Researchers at the University of California: Sharon Begley, "Your Child's Brain," *Newsweek,* 19 February 1996, pp. 55–62.

185 learning actively—that is, through the body: E. Altenmuller, "Cortical DC-Potentials as Electrophysiological Correlates of Higher Hemispheric Dominance of Higher Cognitive Functions," *International Journal of Neuroscience* (47), 1989, pp. 1–14. Cited in N. M. Weinberger, "The Neurobiology of Musical Learning and Memory," *MuSICA Research Notes* 4(2), pp. 5–7.

185 A study in Hong Kong: Agnes S. Chang, et al., "Music Training Improves Verbal Memory," reported in *Nature,* November 1998, p. 128.

185 taking piano lessons and solving math puzzles: Gordon Shaw, *Neurological Research,* 15 March 1999, cited in "Piano and Computer Training Boost Student Math Achievement, UC Irvine Study Shows," *AMC Music-News* (on-line), 15 March 1999.

185 music education majors had the highest reading scores: Don Campbell, *The Mozart Effect* (New York: Avon Books, 1997), p. 177.

185 The College Entrance Examination Board: "Profiles of SAT and Achievement Test Takers 1998," The College Board.

185 musicians' brains are literally different: G. Schlaug, "In Vivo Evidence of Structural Brain Asymmetry in Musicians," *Science,* 267, pp. 699–701.

186 other subjects such as math and science: Lois Birkenshaw-Fleming, "Music for Young Children: Teaching the Fullest Development of Every Child," *Early Childhood Connections,* Spring 1997, pp. 6–13.

CHAPTER 9: MOZART JR.

191 often highly positive, concrete descriptions: Herbert W. Marsh, Rhonda Craven, and Raymond Debus, "Structure, Stability, and Development of Young Children's Self-Concepts: A Multicohort-Multioccasion Study," *Child Development* 69(4), August 1998, pp. 1030–53.

193 only groups who tended to break down music: Dina Kirnarskaya and Ellen Winner, "Musical Ability in a New Key: Exploring the Expressive Ear for Music," unpublished research report, Harvard University and Russian Academy of Music, June 1998, p. 2.

195 By connecting sound, movement, speech, and interaction: David La-

zear, *44 Intelligence Builders for Every Student* (Tucson, Ariz.: Zephyr, 1997).

199 Carolyn Dondero, a certified reading specialist: Carolyn Dondero, "Turning In Teenage Minds with Tunes," unpublished research project, UCLA.

200 Paul Madaule, director of the Listening Centre: Paul Madaule, *When Listening Comes Alive* (Norval, Ontario: Moulin Publishing, 1993).

203 "Several years ago, my youngest . . .": Kerry Hart, "Is It the Melody or the Lyrics?: More Evidence Points to the Power of Music in Learning," *Colorado Music Educator,* Winter 1998, 45(2), pp. 24–25.

204 helped schools in the Saint Paul area: Robert Cutietta, "The Saint Paul Chamber Orchestra CONNECT Program: 1996–1997 Assessment," 5 August 1997, School of Music and Dance, University of Arizona, Tucson.

204 home environment plays an equal: Adele Eskeles Gottfried, James S. Fleming, and Allen W. Gottfried, "Role of Cognitively Stimulating Home Environment in Children's Academic Intrinsic Motivation: A Longitudinal Study," *Child Development* 69(5), October 1998, pp. 1448–60.

205 time to stop or change the pace: Elizabeth Ann McAnally, "Reward Time Can Be Learning Time," *Teaching Music,* June 1998, pp. 34–35.

210 basic principles of Dalcroze Eurhythmics: Virginia Hoge Mead, "More than Mere Movement: Dalcroze Eurhythmics," *Music Educators Journal,* January 1996, pp. 38–37.

POSTLUDE: AN UNFINISHED SYMPHONY

214 *When I was five*: Judith Morley, "Miss Laughinghouse and the Listener," in *Miss Laughinghouse and the Reluctant Mystic* (New York: Black Thistle Press, 1995).

215 "For a few moments music makes us . . .": Robert Jourdain, *Music, the Brain and Ecstasy* (New York: William Morrow & Co., 1997), quoted in *MuSICA Research Notes* 3(2), Fall 1996, p. 1.

RECOMMENDED READING

Abramson, Robert M. *Rhythm Games Book I.* New York: Music and Movement Press, 1973.

———. *Rhythm Games for Perception and Cognition.* Miami, Florida: Volkwein Bros., Inc., 1997.

Benzwie, Teresa. *A Moving Experience: Dance and Learn with Children.* Tucson: Zephyr Press, 1987.

Bérard, Guy, M.D. *Hearing Equals Behavior.* New Canaan, Connecticut: Keats Publishing, 1993.

Bjorkvold, Jon-Roar. *The Muse Within: Creativity and Communication, Song and Play from Childhood through Maturity.* New York: HarperCollins, 1989.

Blood, Peter, and Annie Patterson (eds.). *Rise Up Singing: The Group Singing Songbook.* Bethlehem, Pennsylvania: The Sing Out Corporation, 1992.

Bredekamp, S., and C. Copple (eds.). *Developmentally Appropriate Practice in Early Childhood Programs* (rev. ed.). Washington, D.C.: National Association for the Education of Young Children, 1997.

Burley-Allen, Madelyn. *Listening: The Forgotten Skill.* New York: John Wiley & Sons, 1982.

Cameron, Julia. *The Artist's Way: A Spiritual Path to Higher Creativity.* New York: Tarcher/Putnam, 1992.

———. *The Vein of Gold.* New York: Putnam, 1996.

Cameron, Lindsley. *The Music of Light.* New York: The Free Press, 1998.

Campbell, Don G. *Introduction to the Musical Brain,* 2nd edition. St. Louis, Missouri: MMB Music Inc., 1983.

———. *Master Teacher: Nadia Boulanger.* Washington, D.C.: Pastoral Press, 1984.

———. *The Mozart Effect.* New York: Avon Books, 1997.

———. *100 Ways to Improve Teaching Using Your Voice and Music.* Tucson: Zephyr Press, 1992.

———. *The Roar of Silence.* Wheaton, Illinois: The Theosophical Publishing House, 1989.

Campbell, Don G. (ed.). *Music: Physician for Times to Come.* Wheaton, Illinois: The Theosophical Publishing House, 1991.

———. *Music and Miracles.* Wheaton, Illinois: Theosophical Publishing House, 1992.

Campbell, Don G., and Chris Brewer. *Rhythms of Learning.* Tucson: Zephyr Press, 1991.

Campbell, Linda, Bruce Campbell, and Dee Dickinson. *Teaching and Learning through Multiple Intelligences.* Boston: Allyn and Bacon, 1999.

Cass-Beggs, Barbara. *Your Baby Needs Music: A Music-Sound Book for Babies up to Two Years Old.* New York: St. Martin's Press, 1978.

Chamberlain, D. *The Mind of Your Newborn Baby.* Berkeley, California: North Atlantic Books, 1998.

Chosky, L., R. M. Abramson, A. E. Gillespie, and D. Woods. *Teaching Music in the Twentieth Century.* Englewood Cliffs, New Jersey: Prentice-Hall, 1986.

Clark, Faith, Ph.D., and Cecil Clark, Ph.D. *Hassle-Free Homework: A Six-Week Plan for Parents and Children to Take the Pain Out of Homework.* New York: Doubleday, 1989.

Crustinger, Carla, and Debra Moore. *ADD Quick Tips: Practical Ways to Manage Attention Deficit Disorder Successfully.* Carrollton, Texas: Brainworks, 1987.

DeBeer, Sara (ed.). *Open Ears: Musical Adventures for a New Generation.* Roslyn, New York: Ellipsis Kids, 1995.

Deliège, Irène, and John Sloboda (eds.). *Musical Beginnings: Origins and Development of Musical Competence.* New York: Oxford University Press, 1996.

DePorter, Bobbi. *Quantum Learning: Unleashing the Genius in You.* New York: Dell Publishing, 1992.

Durrell, Doris, Ph.D. *Starting Out Right: Essential Parenting Skills for Your Child's First Seven Years.* Oakland, California: New Harbinger Publications, 1989.

Elkind, David. *Miseducation: Preschoolers at Risk.* New York: Alfred A. Knopf, 1987.

Elliott, David J. *Music Matters.* New York: Oxford University Press, 1985.

Erdei, Peter (ed.). *150 American Folk Songs to Sing, Read, and Play.* New York: Boosey and Hawkes, 1974.

Feierabend, John M. *Music for Little People.* New York: Boosey and Hawkes, 1989.

———. *Music for Very Little People.* New York: Boosey and Hawkes, 1986.

Fowke, Edith. *Sally Go Round the Sun.* New York: Doubleday, 1969.

Gardner, Howard. *Frames of Mind.* New York: Basic Books, 1983.

Goleman, Daniel. *Emotional Intelligence.* New York: Bantam Books, 1995.

Gopnik, Alison, Ph.D., Andrew N. Meltzoff, Ph.D., and Patricia K. Kuhl, Ph.D. *The Scientist in the Crib: Minds, Brains, and How Children Learn.* New York: William Morrow & Co., 1999.

Gordon, Edwin E. *A Music Learning Theory for Newborn and Young Children.* Chicago: GIA Publications, Inc., 1997.

Gordon, Edwin E., and David Woods. *Jump Right In!* Chicago: GIA Publications, 1986.

Gordon, Jay, M.D., and Brenda Adderly, M.H.A. *Brighter Baby.* Washington, D.C.: Lifeline Press, 1999.

Green, Barry. *The Inner Game of Music.* Garden City, New York: Anchor Press, 1986.

Guilmartin, Kenneth K., and Lili M. Levinowitz. *Music and Your Child: A Guide for Parents and Caregivers.* Princeton, New Jersey: Music and Movement Center, 1992.

Habermayer, Sharlene. *Good Music, Brighter Children.* Rocklin, California: Prima Publishing, 1999.

Haines, B. Joan, and Gerber, Linda L. *Leading Young Children to Music.* Columbus, Ohio: Charles E. Merrill Publishing Co., 1980.

Hale, Susan E. *Song and Silence.* Albuquerque: La Alameda Press, 1995.

Hannaford, Carla, Ph.D. *The Dominance Factor: How Knowing Your Dominant Eye, Ear, Brain, Hand and Foot Can Improve Your Learning.* Arlington, Virginia: Great Ocean Publishers, 1997.

———. *Smart Moves: Why Learning Is Not All In Your Head.* Arlington, Virginia: Great Ocean Publishers, 1995.

Harris, Robert. *What to Listen for in Mozart.* New York: Simon & Schuster, 1991.

Hayden, Robert C. *Singing for All People: Roland Hayes, a Biography.* Boston: Select Publications, 1989.

Hildesheimer, Wolfgang. *Mozart.* New York: Farrar Straus & Giroux, 1982.

Hoffman, Janalea. *Rhythmic Medicine.* Leawood, Kansas: Jamillan Press, 1995.

Janov, A. *Imprints: The Lifelong Effects of the Birth Experience.* New York: Coward-McCann, Inc., 1983.

Janus, L. *The Enduring Effects of Prenatal Experience.* Princeton, New Jersey: Jason Aronson, Inc., 1997.

Jarnow, Jill. *All Ears: How to Choose and Use Recorded Music for Children.* New York: Penguin Books, 1991.

Jenkins, Peggy, Ph.D. *The Joyful Child.* Santa Rosa, California: Aslan Publishing, 1996.

Jones, Bessie, and Bess Hawes. *Step It Down: Games, Plays, Songs, and Stories from the Afro-American Heritage.* New York: Harper and Row, 1972.

Judy, Stephanie. *Making Music for the Joy of It: Enhancing Creativity, Skills, and Musical Confidence.* Los Angeles: Jeremy P. Tarcher, Inc., 1990.

Kaner, Etta. *Sound Science.* Reading, Massachusetts: Addison-Wesley, 1991.

Kaplan, Don. *See with Your Ears: The Creative Music Book.* San Francisco: Lexicos, 1983.

Kline, Peter. *The Everyday Genius: Restoring Children's Natural Joy of Learning—and Yours Too.* Arlington, Virginia: Great Ocean Publishers, 1988.

Kodály, Zoltán. *Fifty Nursery Songs.* New York: Boosey and Hawkes, 1964.

Kristel, Dru. *Breath Was the First Drummer.* Santa Fe, New Mexico: QX Publications A.D.A.M., Inc., 1995.

Lafuente, M. J., et al. "Effects of the Firstart Method of Prenatal Stimulation on Psychomotor Development: The First Six Months." *Prenatal and Perinatal Psychology Journal* 11(3), pp. 151–62.

Landalf, Helen, and Pamela Gerke. *Movement Stories.* Lyme, New Hampshire: Smith and Kraus, Inc., 1996.

Lane, Deforia. *Music as Medicine.* Grand Rapids, Michigan: Zondervan Publishing House, 1994.

Lazear, David. *44 Intelligence Builders for Every Student.* Tucson: Zephyr Press, 1997.

———. *Seven Pathways of Learning: Teaching Students and Parents about Multiple Intelligences.* Tucson: Zephyr Press, 1994.

Leach, Penelope. *Your Baby and Child: From Birth to Age Five.* New York: Alfred A. Knopf, 1980.

Lynch, Stacy Combs. *Classical Music for Beginners.* New York: Writers and Readers Publishing, Inc., 1994.

Machover, W., and M. Uszler. *Sound Choices: Guiding Your Child's Musical Experiences.* New York: Oxford University Press, 1996.

Madaule, Paul. *When Listening Comes Alive.* Norval, Ontario: Moulin Publishing, 1993.

Marks, Kate (ed.). *Circle of Song: Songs, Chants, and Dances for Ritual and Celebration.* Lenox, Massachusetts: Full Circle Press, 1993.

Marshall, Robert L. *Mozart Speaks.* New York: Schirmer Books, 1991.

McDonald, Dorothy T., and Gene M. Simons. *Musical Growth and Development, Birth through Six.* New York: Schirmer Books, 1989.

Meltzer, D. (ed.). *Birth: An Anthology of Ancient Texts, Songs, Prayers, and Stories.* San Francisco: North Point Press, 1981.

Mike, John, M.D. *Brilliant Babies, Powerful Adults.* Clearwater, Florida: Satori Press International, 1997.

Mills, E., and Sr. Therese Cecile Murphy. *The Suzuki Concept: An Introduction to a Successful Method for Early Music Education.* Berkeley, California: Diablo Press, Inc., 1973.

Newham, Paul. *The Singing Cure: An Introduction to Voice Movement Therapy.* Boston: Shambhala Publications, Inc., 1993.

Nordhoff, P., and C. Robbins. *Therapy in Music for Handicapped Children,* 3rd ed. London: Victor Golancz, 1992.

Ostrander, S., and L. Schroeder. *Superlearning 2000.* New York: Delacorte Press, 1992.

Page, Nick. *Music as a Way of Knowing.* York, Maine: Stenhouse Publishers, 1996.

Paley, Vivian Gussin. *Mollie Is Three: Growing Up in School.* Chicago: University of Chicago Press, 1986.

Palmer, Mary, and Wendy L. Sims (eds.). *Music in Prekindergarten.* Reston, Virginia: Music Educators National Conference, 1993.

Papoušek, M. "Early Ontogeny of Vocal Communication in Parent-Infant Interactions," in H. Papoušek, U. Jurgens, and M. Papoušek (eds.), *Nonverbal Vocal Communication.* Cambridge: Cambridge University Press, 1993.

Parouty, Michel. *Mozart: From Child Prodigy to Tragic Hero.* New York: Harry N. Abrams, 1993.

Pearce, Joseph Chilton. *Evolution's End: Claiming the Potential of Our Intelligence.* New York: HarperCollins, 1992.

Pearce, Joseph Chilton. *Magical Child.* New York: Plume, 1977.

Ristad, Eloise. *A Soprano on Her Head.* Moab, Utah: Real People Press, 1982.

Rose, Colin. *Accelerated Learning.* Suffolk, England: Accelerated Learning Systems Ltd., 1985.

Sabbeth, Alex. *Rubber-Band Banjos and a Java Jive Bass: Projects and Activities on the Science of Music and Sound.* New York: John Wiley & Sons, 1997.

Sale, Laurie. *Growing Up with Music: A Guide to the Best Recorded Music for Children.* New York: Avon Books, 1992.

Schafer, R. Murray. *Creative Music Education: A Handbook for the Modern Music Teacher.* New York: Macmillan, 1976.

Schiller, Pam, and Thomas Moore. *Where Is Thumbkin?* Beltsville, Maryland: Gryphon House, 1993.

Schuyler, Valerie, M.A., Jayne Sowers, Ed.D., and Norene Browles. *Parent-Infant Communications* (4th Ed.), and *For Families Guidebook.* Hearing & Speech Institute, 1998. Two one-hour videos available separately or in a complete package. Contact at (503) 228-6479 or at the Web site: Valeries@hearingandspeech.org.

Seeger, Ruth. *American Folk Songs for Children.* New York: Doubleday, 1948.

Shaw, Gordon L., Ph.D. *Keeping Mozart in Mind.* San Diego: Academic Press, 1999.

Silberg, Jackie. *Games to Play with Babies.* Beltsville, Maryland: Gryphon House, 1993.

Solomon, Maynard. *Mozart: A Life.* New York: HarperCollins, 1995.

Stern, Daniel N., M.D. *Diary of a Baby: What Your Child Sees, Feels, and Experiences.* New York: Basic Books, 1990.

Storms, Jerry. *101 Music Games for Children.* Alameda, California: Hunter House, Inc., 1995.

Stwertka, Eve, and Albert Stwertka. *Tuning In the Sounds of the Radio.* New York: Julian Messner, 1992.

Taylor, Jack A., Nancy H. Barry, and Kimberly C. Walls. *Music and Students at Risk: Creative Solutions for a National Dilemma.* Reston, Virginia: New Music Educators National Conference, 1997.

Tomatis, Alfred A., M.D. *The Conscious Ear.* Barrytown, New York: Station Hill Press, 1991.

———. *The Ear and Language.* Norval, Ontario: Moulin Publishing, 1996.

———. *Ecouter l'Univers.* Paris: Editions Robert Laffont, 1996.

———. *La Nuit Uterine.* Paris: Editions Stock, 1981.

———. *Nous Sommes Tous Nes Polyglottes.* Paris: Editions Pixot, 1991.

———. *L'Oreille et le Voix.* Paris: Editions Robert Laffont, 1987.

———. *Pourquoi Mozart?* Paris: Editions Fixot, 1991.

Trehub, S. E., and L. J. Trainor. "Listening Strategies in Infancy: The Roots of Music in Language Development." In S. McAdams, E. Bigand, et al. (eds.). *Thinking in Sound: The Cognitive Psychology of Human Audiation.*

Oxford, England: Clarendon Press/Oxford University Press, 1993, pp. 278–327.

Verny, Thomas, M.D., with John Kelly. *The Secret Life of the Unborn Child: How You Can Prepare Your Unborn Baby for a Happy, Healthy Life.* New York: Dell Publishing, 1981.

Verny, Thomas, M.D., with Pamela Weintraub. *Nurturing the Unborn Child.* New York: Delacorte, 1991.

Wegner, Win, Ph.D., and Richard Poe. *The Einstein Factor.* Rocklin, California: Prima Publishing, 1996.

Weikart, Phyllis S. *The Round Circle: New Experiences in Movement for Children.* Ypsilanti, Michigan: The High/Scope Press, 1987.

Whitfield, Charles L., M.D. *Healing the Child Within.* Deerfield Beach, Florida: Health Communications, Inc., 1987.

Wirth, M., V. Stassevitch, R. Shotwell, and P. Stemmier. *Musical Games, Fingerplays and Rhythmic Activities for Early Childhood.* West Nyack, New York: Parker Publishing Company, 1983.

Wiseman, Ann. *Making Musical Things.* New York: Charles Scribner's Sons, 1979.

Wolvin, A. D., and C. G. Coakley (eds.). *Perspectives on Listening.* Norwood, New Jersey: Ablex Publishing Corporation, 1993.

AUDIO

Coulter, Dee J. *Beginning with Music: A Guided Tour of the Brain; Meeting the World with Music; Music's Gift for a Developing Mind.* Kindling Touch Publications, 4850 Niwot, Longmont, Colorado 80503, (303) 530-2357.

Cry-No-More™. Smart Baby™ Care, 1996.

Fay, Jim, and Foster W. Cline, M.D. *Toddlers: Love and Logic Parenting for Early Childhood.* Golden, Colorado: Love and Logic Press, Inc., 1997.

Koomar, Jane, Ph.D., and Stavey Szklut, M.S. *Making Sense of Sensory Integration.* Boulder, Colorado: Bell Curve Records, 1998.

Olkin, Sylvia K. *Relax and Enjoy Your Baby.* Roslyn, New York: The Relaxation Company, Inc., 1996.

Sound Beginnings: Languages and Music of the World. Dallas, Texas: Sound Beginnings, 1998.

THE MUSIC OF MOZART*

FOR PREGNANCY

- *Variations on Ah! Vous dirai-je, Maman* (K. 265). Whether they start you singing "Twinkle Twinkle, Little Star," "Baa, Baa Black Sheep," or the French folk song on which they were based, these sparkling variations will stimulate your growing baby's brain development—and cheer you both up at the same time.
- The Andantino from the Flute Quartet in C Major (K. 171) (258b). In this playful excerpt from the flute quartet, Mozart almost tells a story. The light, active, and fresh music is easy on the ears, excellent "ear candy" for your developing baby.
- Andante from the Symphony No. 25 in G Minor (K. 183). In this symphony, Mozart uses a perfect "Go to Sleep" theme. Listen carefully and you will hear the music saying "Go to Sleep" in its melody. On nights when your baby refuses to settle down, try playing this for her as you sing or chant "Go to Sleep" every time you hear the theme.

FOR INFANTS

- *Variations on Ah! Vous dirai-je, Maman* (K. 265). If you played these variations on "Twinkle Twinkle, Little Star" for your baby before his birth, they are sure to remain one of his favorites. Hold and rock your infant while you listen to the music together, recalling a time when the two of you were one. You might sing "The Alphabet Song," "Baa, Baa Black Sheep," or make up your own words to go with the melody.
- Andante from the Symphony No. 25 in G Minor (K. 183). Again, this selection from one of Mozart's symphonies will sound wonderfully

*Available on *The Mozart Effect for Children* and *The Mozart Effect for Babies* CD and cassette series.

familiar to your infant if you played it before his birth. This is not a traditional lullaby but rather a piece of music that invites you to speak, sing, or chant while holding your little one close. After a few minutes, your newborn will be ready for more restful music.

- Andante Sostenuto from the Violin Sonata in C Major (K. 296). Now the feeling of a lullaby comes to soothe you and your baby. Allow the stress and structure of the day to melt away as the music balances mind, heart, and body. Cradle your baby close to you. Can you feel your newborn respond to the physical changes in your body?

FOR BABIES

- The Minuet from *The Toy Symphony* by Leopold Mozart. This is the delightful piece composed by Mozart's father a few months before Wolfgang was born. The "little cuckoo," horn, and glockenspiel call out a delightful tune, inviting you to play "pat-a-cake" or "peekaboo" with your baby. You'll hear, too, a rubber ducky and bird join in the musical celebration. This piece is perfect for engaging playtime fun.
- German Dance No. 2 (K. 605). Have your child move her body or stand and sway, allowing the magic of these great German dances to become a part of her play and dance life.
- The Rondo—Allegro ma non troppo from the Serenade No. 9 in D Major (K. 320). A rondo repeats a melody section of music with variations in between. Use this selection to dance, gently rock, or quietly play with your baby. You're sure to enjoy this quieter time, too!
- The Andantino from the Flute Quartet in C Major (K. 171) (258b). This quiet and lovely piece is relaxing for everyone.
- The Adagio III from the Quartet No. 20 in D Major (K. 499). Turn the music down softly before you leave the baby's room. The quiet music will mask the other sounds in the house and slowly quiet the mind and body of your baby.
- *Variations on Ah! Vous dirai-je Maman* (K. 265) as played on the organ. We return to a soothing, slow variation of "Twinkle Twinkle, Little Star" on the organ. The low sounds offer a quiet background to turn out the lights, say good night, and pray your baby has a long, undisturbed night's sleep.

FOR TODDLERS

- The Adagio from the Divertimento in B Flat (K. 287). The Italian word *adagio* means "to put at ease." This music is slow and leisurely, perfect for an afternoon nap. Invite your child to close his eyes and let his body rest as Mozart takes him on a relaxed journey through sound.
- March No. 1 (K. 335). Here is a fine march for preparing your child to get dressed and go outside. Its lively beat will put mind and body in motion and in order, and inspire your little one to move.
- "Champagne Aria" from *Don Giovanni.* Dance. Skip. Twist. Turn. Invent wild motions. Dance along with Mozart's wonderful aria from *Don Giovanni.* You can dance along with your child or use this for your own aerobic exercise.

FOR PRESCHOOLERS

- The Concertante from the Serenade No. 9 in D Major (K. 320). Mozart composed serenades for gala events in royal palaces. This music was used as a kind of background for parties and meeting people. Help your child imagine a fine procession of people dressed up in their best for a great party in the wigs, great hoop skirts, and fancy attire that Mozart must have seen as she listens to this wonderful concertante.
- *Rondo à la Turke* (K. 331). This famous rondo is a great piece for clapping along, moving to the music, and listening for the accents in the rhythm. A great workout for mind and body.
- The Catillion and the Allegro from the Divertimento in B Flat (K. 15). This is a perfect hide-and-seek piece. Look around, find a friend, run and hide to the music, and then sit down and play it again while you rest.
- "Papageno's Song" from *The Magic Flute* (K. 620). Help your child imagine playing a magic flute as she skips, dances, and twirls—bringing the magic of Mozart's beauty to you and to the ears of her friends.

FOR KINDERGARTNERS

- The Rondo from the *Eine Kleine Nachtmusik* (K. 525). This rondo is one of the most charming melodies ever written and easily brings a child's mind to attention. Use this music to alert your child's mind that desk time is approaching, and that it can be joyous and happy.

- "Vol che sapete" from *The Marriage of Figaro.* This haunting and beautiful song from *The Marriage of Figaro* can easily be translated as, "You have the answer, you have the key." Its beautiful melody allows your child to fantasize about beauty and joy.
- The Andante from the Cassation (K. 63). *Andante* means "walking slowly." An andante is never too fast, and gives us a sense of walking or slow movement. Suggest that your child close his eyes as he listens and imagine what it was like when Mozart, a young child of eight or nine, heard this melody in his head and wrote it down. Perhaps he was traveling in a carriage from one city to another and was inspired by what he saw out the window.
- *Variations on Ah! Vous dirai-je, Maman* (K. 265). Learning to approach desk work with variety and freshness is important. It's best to have a regular routine and enter into it every day with a fresh new mind. Just as this lovely piece stimulated your infant's mind, so it can awaken his brain to new ideas now as he begins school.
- *Die Leyerer* (K. 611). Ask your child to imagine himself walking with young Mozart through one of the beautiful cities of Europe. The two of them come across a hurdy-gurdy man playing his primitive instrument. Encourage your child to listen, dance, and sway to these beautiful and unique sounds.
- *La Bataille* (K. 535). Here is a whole story of a little battle between tin soldiers. Give your child the freedom to make up his own story, play, or pantomime. It's all in the music.
- The Menuetto from the Divertimento in B Flat (K. 287). Minuets are dances that often have a waltzlike beat. Here's your child's opportunity to make up his own ballet, exercise, or stretching movements. Encourage him to use his hands, feet, elbows, knees, and nose while seeing how many ways he can put Mozart into motion.

For Early Elementary School

- Variations from the Sinfonia (K. 297b). Repetition is one of the best-known ways to learn information. When we repeat information or musical melodies in a variety of ways, the brain and the ear begin to listen differently. Ask your child to listen to this piece and see if she can tell what melody is repeated all the way through the piece. If she feels drowsy, invite her to stretch and gently move to this music to help the mind and the body prepare itself for study and concentration.

- The Andante from the Symphony No. 6 (K. 43). Drawing pictures can help activate the right brain's natural spatial development. Offer your child paper and chalk or crayons, and allow the sweeping sounds of this andante from Symphony No. 6 to inspire her imagination.
- The Andantino grazioso from the Symphony No. 18 (K. 130). When beginning to feel the stress of study or exercise, this slow, magical piece can help your child relax. Put on the CD, suggest that she close her eyes, and let Mozart give her ears a gentle, relaxing massage.
- The Andantino from the Symphony No. 24 (K. 182). The Andantino from Symphony No. 24 is one of the most perfect pieces that Mozart ever wrote for bringing language to music. Ask your child to close her eyes as she listens to this piece, and see if she can hear the story it's telling. The strings and the flute are very clearly repeating the sound information and speak in a clear, direct form. By listening to the ways in which music develops an idea, asks questions, and then allows the listener to come to resolution, your child is exposed to one of the best ways of writing an essay or a speech.
- The Allegro aperto from the Violin Concerto No. 5 in A Major (K. 219). This sparkling allegro can help charge the brain and give your child's ears a good deal of exercise. Have her spend ten minutes before her study time imagining that she is conducting the orchestra. This process of active listening and moving will bring her brain and body into full attention, preparing her for better concentration.
- The Prestissimo from the Serenade in D Major (K. 203). This prestissimo, which means "very fast," can allow your child to develop her own production of a ballet, including many characters with many different feelings. Suggest that she listen to this piece many times with her eyes closed, watching a ballet, seeing what happens to the characters, and letting her imagination and her body move together.

FOR LATE ELEMENTARY SCHOOL

- The Allegro moderato from the Violin Concerto No. 2 in D Major (K. 211). This violin concerto has been used to help tens of thousands of children retune their ears at over two hundred Tomatis-inspired

Listening Centers throughout the world. Dr. Tomatis would suggest that your child listen to this piece with his right ear focused toward the stereo speaker, thus stimulating his language centers before studying.

- The Andante from the Symphony No. 17 (K. 120). This is a good piece to listen to before study if your child is already energized. The high frequency and the slower tempo allow him to slow down and yet be stimulated at the same time.
- The Adagio and Gran Partita from the Serenade No. 10 in B Flat (K. 361). The Gran Partita is one of Mozart's most elegant and brilliant pieces. Have your child close his eyes and imagine he is in the courtyards of the royal courts of Salzburg or Vienna. As the music plays, he can watch the story unfold.
- The Andante from the Symphony No. 15 (K. 124). This andante, which Mozart wrote when he was young, is a wonderful selection to begin your child's study session as he sits down and looks at the materials in front of him. What does he need to do? What does he need to read? How long does he have? While listening to this music, he can organize his desk, his ear, and his mind.

CHILDREN'S SONGS

For Pregnancy

This Little Light of Mine
Twinkle Twinkle, Little Star
He's Got the Whole World
Swing Low, Sweet Chariot
The Water Is Wide
You Are My Sunshine
Rock My Soul
Turn, Turn, Turn

For Infants

It's Raining, It's Pouring
Twinkle Twinkle, Little Star
Lullaby and Good Night
All the Pretty Little Horses
Oh How Lovely Is the Evening
Baa, Baa Black Sheep
Kumbaya
A-B-C-D-E-F-G
Mary Had a Little Lamb
Five Little Monkeys
There Were Five in the Bed
Muffin Man
Tumbalalaika
My Little Pony
Hey Diddle Diddle
Rain, Rain, Go Away
One, Two, Buckle My Shoe

For Babies

Row, Row, Row Your Boat
Hot Cross Buns
Pat-a-Cake
Jack and Jill
A Tisket, a Tasket
I'm a Little Teapot
Pease Porridge Hot
Humpty-Dumpty
This Little Piggy
All Through the Night
Where Is Thumbkin?
Ride a Cock Horse
Old King Cole
To Market, to Market
Baa, Baa Black Sheep
Old MacDonald

FOR TODDLERS

Oats, Peas, Beans
Head and Shoulders, Knees and Toes
There's a Little White Duck
There Were Ten in the Bed
Pop Goes the Weasel
Polly Put the Kettle On
Ring around the Rosey
Biscuits in the Oven
Down at the Station
Did You Ever See a Lassie
Take Me Out to the Ball Game
Five Little Ducks
Mulberry Bush
Yankee Doodle
Riding in My Car
Frère Jacques
London Bridge
Clap, Clap, Clap Your Hands
If You're Happy and You Know It
Put Your Finger in the Air
One, Two, Buckle My Shoe
I'm a Little Teapot
This Old Man
Shortnin' Bread
Alphabet Song
Rig-a-Jig-Jig
Skip to My Lou

FOR PRESCHOOLERS

Here We Go Looby-Loo
Shake My Sillies Out
Ha Ha This-a-Way
Get on Board Little Children
Sweetly Sings the Donkey
The More We Get Together
Where, Oh Where Has My Little Dog Gone?
The Hokey Pokey
Little Sir Echo
Peanut Butter
Michael Finnigan
This Train
Go In and Out the Window
The Little Nut Tree
Old Brass Wagon
Apples and Bananas
Magic Penny
Love Somebody

FOR KINDERGARTNERS

Going to the Zoo
Down by the Bay
Polly Wolly Doodle
The Bear Went over the Mountain
She'll Be Comin' Round the Mountain
Take Me Out to the Ball Game
I've Been to London
B-I-N-G-O
A Hunting We Will Go
I've Been Working on the Railroad
Doe, a Deer (The "Do Re Mi" Song)
A Ram Sam Sam

Who Stole the Cookie from the Cookie Jar?
Barnyard Dance
Kookaburra
A Sailor Went to Sea

FOR EARLY ELEMENTARY SCHOOL

On Top of Spaghetti
Valderi Valdera
Over the River
Be Kind to Your Web-Footed Friends
Michael Row Your Boat Ashore
There's a Hole in the Bucket
Boom Boom Ain't It Great to Be Funny
Day-O
Dem Bones
My Favorite Things
Home on the Range
Cockles and Mussels
Scotland's Burning
One Meatball
Purple People Eater
I've Got Sixpence
The Garden Song
Heigh-Ho
The Marvelous Toy
Li'l Liza Jane
Froggie Went a Courtin'
I Got Shoes
This Land Is Your Land
Erie Canal

FOR LATE ELEMENTARY SCHOOL

Over the Rainbow
Edelweiss
Oh Susannah
Puff the Magic Dragon
Amazing Grace
Scarborough Fair
Where Have All the Flowers Gone?
Swing Low, Sweet Chariot
Fish and Chips and Vinegar
Down in the Valley
The Cat Came Back
Everybody Loves Saturday Night
Yellow Submarine
On Top of Old Smoky
Down by the Riverside
All Night, All Day
Freight Train
Simple Gifts
Little Boxes

SPANISH LANGUAGE SONGS FOR CHILDREN

Pin Pon
El Coquí
La Serena
Kikiriki

Mi Chacra
Los Pollitos
De Colores
Donde Están las Llavas
Arroyo Clara
Las Mananitas
Un Elefante
La Cucaracha
La Bamba
Guantanamera
Mi Burro
Cabalito Blanco

INDEX

Books by Don Campbell:

THE MOZART EFFECT
Tapping the Power of Music to Heal the Body,
Strengthen the Mind, and Unlock the Creative Spirit
ISBN 0-06-093720-3 (paperback)

Drawing on medicine, Eastern wisdom, and the latest research on learning and creativity, Campbell reveals how exposure to sound, music and other forms of vibration can have a lifelong effect on health, learning and behavior. He shows how to use sound and music to simulate learning and memory; how to strengthen listening abilities; how to use imagery to enhance the Mozart Effect; and how to harness the power of toning, chanting, mantras, rap and other self-generated sounds. Listing over fifty common conditions for which music can be used as a treatment or cure, Campbell then recommends more than two dozen specific, easy-to-follow exercises to help you raise your spatial IQ, sound away pain, boost creativity, and make the spirit sing.

"Campbell's work is of inestimable value. Practical, mystical and visionary, he makes the world of music accessible, friendly, and profoundly healing."
—Julia Cameron, author of *The Artist's Way*

THE MOZART EFFECT FOR CHILDREN
Awakening Your Child's Mind, Health and Creativity With Music
ISBN 0-380-80744-0 (paperback)

Campbell presents to parents and teachers the first specific program that uses music to enhance the life of children through rhythm and tone. In this book, Campbell follows a child's life from pre-birth through age 10, demonstrating ways in which music can be used to activate new neural pathways in the brain of the fetus and infants, promote language acquisition, prepare the brain for reading, and much more. The book includes music and movement exercises to stimulate children's minds, music "recipes" for helping children internalize a rhythm of thought and mental organization, and recommendations for reducing stress and maintaining family cohesiveness through music.

"A bountiful compendium of research, teaching, and learning strategies, and innumerable other resources that enrich and stretch the human mind, body, and spirit to new dimensions."—Dee Dickinson, CEO, New Horizons for Learning

Available wherever books are sold, or call 1-800-331-3761 to order.

Don Campbell has produced numerous sets of recordings of Mozart's music to help parents interested in taking advantage of the Mozart Effect in their own homes. Each set is available on both compact disc and cassette.

Music for the Mozart Effect®

Volume I—Strengthen the Mind:
Music for Intelligence and Learning
Volume II—Heal the Body:
Music for Rest and Relaxation
Volume III—Unlock the Creative Spirit:
Music for Creativity and Imagination
Volume IV—Music for Stress Reduction
Volume V—Music for Study

Available from Springhill Music (800) 427-7680
www.springhillmedia.com

The Mozart Effect®—Music for Children

Volume I—Tune Up Your Mind
Volume II—Relax, Daydream, and Draw
Volume III—Mozart in Motion
Volume IV—Mozart to Go!

The Mozart Effect®—Music for Babies

Volume I—From Playtime to Sleepytime
Volume II—Music for Newborns
Volume III—Nighty-Night Mozart
The Mozart Effect®—Music for Moms and Moms-to-Be

Available from The Children's Group (800) 757-8372
www.childrensgroup.com